WRITING IN POLITICAL SCIENCE

BRIEF GUIDES TO
WRITING IN THE DISCIPLINES

EDITED BY
THOMAS DEANS, *University of Connecticut*
MYA POE, *Northeastern University*

Although writing-intensive courses across the disciplines are now common at many colleges and universities, few books meet the precise needs of those offerings. These books do. Compact, candid, and practical, the *Brief Guides to Writing in the Disciplines* deliver experience-tested lessons and essential writing resources for those navigating fields ranging from biology and engineering to music and political science.

Authored by experts in the field who also have knack for teaching, these books introduce students to discipline-specific writing habits that seem natural to insiders but still register as opaque to those new to a major or to specialized research. Each volume offers key writing strategies backed by crisp explanations and examples; each anticipates the missteps that even bright newcomers to a specialized discourse typically make; and each addresses the irksome details that faculty get tired of marking up in student papers.

For faculty accustomed to teaching their own subject matter but not writing, these books provide a handy vocabulary for communicating what good academic writing is and how to achieve it. Most of us learn to write through trial and error, often over many years, but struggle to impart those habits of thinking and writing to our students. The *Brief Guides to Writing in the Disciplines* make both the central lessons and the field-specific subtleties of writing explicit and accessible.

These versatile books will be immediately useful for writing-intensive courses but should also prove an ongoing resource for students as they move through more advanced courses, on to capstone research experiences, and even into their graduate studies and careers.

OTHER AVAILABLE TITLES IN THIS SERIES INCLUDE:

Writing in Engineering: *A Brief Guide*

Robert Irish
(ISBN: 9780199343553)

Writing in Biology: *A Brief Guide*

Leslie Ann Roldan and Mary-Lou Pardue
(ISBN: 9780199342716)

WRITING IN POLITICAL SCIENCE

A BRIEF GUIDE

Mika LaVaque-Manty
Danielle LaVaque-Manty
UNIVERSITY OF MICHIGAN

Oxford University Press
New York

Oxford University Press is a department of the University of Oxford.
It furthers the University's objective of excellence in research,
scholarship, and education by publishing worldwide.

Oxford   New York
Auckland   Cape Town   Dar es Salaam   Hong Kong   Karachi
Kuala Lumpur   Madrid   Melbourne   Mexico City   Nairobi
New Delhi   Shanghai   Taipei   Toronto

With offices in
Argentina   Austria   Brazil   Chile   Czech Republic   France   Greece
Guatemala   Hungary   Italy   Japan   Poland   Portugal   Singapore
South Korea   Switzerland   Thailand   Turkey   Ukraine   Vietnam

For titles covered by Section 112 of the US Higher Education
Opportunity Act, please visit www.oup.com/us/he for the latest
information about pricing and alternate formats.

Published by Oxford University Press
198 Madison Avenue, New York, New York 10016
http://www.oup.com

Library of Congress Cataloging-in-Publication Data

Names: LaVaque-Manty, Danielle, 1968-, author. | LaVaque-Manty, Mika,
    1966-, author.
Title: Writing in political science : a brief guide / Danielle LaVaque-Manty
    and Mika LaVaque-Manty, University of Michigan.
Description: New York, New York : Oxford University Press, [2015] | Series:
    Writing in the disciplines | Includes bibliographical references.
Identifiers: LCCN 2015020337 | ISBN 9780190203931 (acid-free paper)
Subjects: LCSH: Political science--Authorship--Style manuals. | Political
    science--Research--Handbooks, manuals, etc. | Academic writing--Handbooks,
    manuals, etc.
Classification: LCC JA86 .L38 2015 | DDC 808.06/632--dc23 LC record available
    at http://lccn.loc.gov/2015020337

Printing number: 9 8 7 6 5 4 3 2 1

Printed in the United States of America
on acid-free paper

*For the students from whom we have learned
so much and for the students whom we hope
this book will help.*

BRIEF TABLE OF CONTENTS

TABLE OF CONTENTS

CHAPTER 4 **Strategies for Data-Driven Research Proposals and IMRD Papers** 73

CHAPTER 5 **Strategies for Response
Papers, Case Studies,
Advocacy Papers, and
Blog Posts** **110**

CHAPTER 6 **Style is Meaning** **123**

PREFACE

What counts as good academic writing varies from one discipline to the next. This can be confusing to students and invisible to faculty members. Students get confused as the rules about writing shift when they move from English class to biology to political science. Faculty may forget the differences when the characteristics of their own fields have become too familiar to be noticeable. The variety in academic writing doesn't have to be a problem: it offers students an opportunity to investigate the disciplinary differences and develop their rhetorical skills. But it's difficult for them to do this on their own, and few faculty have time or resources available to help them. That's where this book comes in: it describes patterns of writing that are specific to political science and explains the logic behind those patterns.

Writing and thinking go hand in hand; thus, our approach in this book is to explain how political scientists think, and connect that to their expectations about how students should write. In many ways, thinking like a political scientist means thinking like a social scientist, but with a focus on certain kinds of questions and preoccupations having to do with actors (voters, rulers, representatives, parties, classes, states) pursuing their ends (power, justice, public goods) in a world shaped by institutions (laws, norms, markets) and competition. One challenge to teaching students to write well in political science is that because the subject matter is political, people have strong and sometimes passionate views about it. Hence, we try to teach writers (1) to identify and operate in different registers and with different levels of analysis, and thereby (2) to develop a disposition of respectful engagement with others' ideas. We try to help them understand that political, normative, and evaluative disagreements don't have to map onto factual, empirical, or explanatory controversies. Whether one is a Republican or a Democrat doesn't determine what one can know about the question of whether democracy causes development or about the effects of priming and agenda setting. In other words, in this book we try to help students learn how to participate in what we call the ongoing "scholarly conversation" that is political science.

In the pages that follow, we explain the rhetorical concepts of audience, purpose, genre, and credibility, applied to the political science context. We then walk our readers through the process of writing a paper: understanding what their writing assignments are asking them to do, generating interesting and answerable research questions, providing appropriate evidence for their arguments, and offering satisfactory answers to the "So what?" question in their conclusions. We have included chapters that pay detailed attention to word choice and style, and to source selection and documentation. The advice we offer applies to papers of varying length and complexity, from response papers to honors theses, and across subfields. We wrote the book primarily for undergraduates who have decided to major in political science, but new graduate students may find it helpful as well, particularly if they did not major in political science in college. Students who are still considering political science—students in introductory courses, for example—will find this another type of introduction to the discipline. We also believe the book's attention to stylistic nuances and expected organizational structures will be useful to students whose first language is not English.

In addition to detailed but concise advice about how to write good political science papers, the book contains several checklists highlighting key points. It also includes an appendix on seeking and using feedback, and another with suggestions for further reading for students working with data they have collected themselves.

ABOUT THE AUTHORS

MIKA LAVAQUE-MANTY is Arthur F. Thurnau Professor and Associate Professor of Political Science at the University of Michigan. He is the author of *Arguments and Fists: Political Agency and Justification in Liberal Theory* (Routledge, 2002) and *The Playing Fields of Eton: Equality and Excellence in Modern Meritocracy* (University of Michigan Press, 2009), which won the University of Michigan Press Award for best book published by the press over a two-year period. He teaches courses in political theory and political science research design, many of which are writing intensive. He has won numerous teaching awards, including, most recently, the Provost's Teaching Innovation Prize in 2013, and he serves as a Faculty Associate at the Center for Research on Learning and Teaching at the University of Michigan.

DANIELLE LAVAQUE-MANTY earned a Ph.D. in political science at the University of Michigan and an MFA in creative writing at The Ohio State University. She is coeditor of *Using Reflection and Metacognition to Improve Student Learning* (Stylus Publishing, 2013) and *Transforming Science and Engineering: Advancing Academic Women* (University of Michigan Press, 2007). She taught academic writing and writing pedagogy at the University of Michigan's Sweetland Center for Writing for five years and currently freelances as an academic editor and writing consultant.

ACKNOWLEDGMENTS

This book, like so much writing in any discipline, is a result of many different kinds of collaboration. You are reading this book—as opposed to a website or a set of blog posts—because of acquisitions and production staffs at Oxford University Press and their partners who brought it about. We are grateful to them, particularly Garon Scott.

We are also profoundly grateful to Tom Deans and Mya Poe, the editors of this series of books on writing in the disciplines. From our initial email exchanges to Skype conversations, they have been ideal editors: they are careful and constructive readers, and supportive throughout. They read and commented on a penultimate draft of the manuscript over four days of holidays during a winter break. That is probably a record of some kind and certainly deserves a karmic prize, as far as we are concerned.

We are also grateful to Michelle Cox for her help in making this book as accessible as possible to students whose first language is not English, and to the many anonymous reviewers whose advice helped improve the manuscript immensely.

Kimberly K. Smith at Carleton College initially put us in touch with Tom and Mya and provided assignments we discuss in the book besides. We have loved to talk, agree, and disagree with Kim for more than twenty years and are, once again, grateful for her help.

Many, many other colleagues both at Michigan and elsewhere provided assignments we discuss in this book, commented on drafts, and, in general, helped make this book better.

Particular thanks go to Jeff Bernstein, Sarah Croco, Tom Flores, Terri Givens, Khristina Haddad, Nathan Kalmoe, Elizabeth Mann, Brian Min, Ben Smith, and Nick Valentino.

Thank you as well to those who reviewed the book in proposal or manuscript form:

Fred Cocozzelli, St. John's University; James Endersby, University of Missouri; Justin Esarey, Rice University; Donna Evans, Eastern Oregon University; Jamie Frueh, Bridgewater University; Alison Gash, University of Oregon; Kenji Hayao, Boston College; Milda Hedblom, Augsburg College; Kim Hill, Texas A&M University; Mirya Holman, Florida Atlantic University; Amy Lannin, University of Missouri; Noreen Lape, Dickinson University; Susan Liebell, Saint Joseph's University; Kimberly Morgan, George Washington University; Richard Niemi, University of Rochester; Zachary Shirkey, Hunter College; Markus Smith, University of Central Oklahoma; James Truman, Auburn University; Frederick Wood, Coastal Carolina University.

We are most grateful to the undergraduate students for whom this book is primarily meant. We both have benefited, indeed learned, from our students, in the preparation of this book and in our jobs. Several of our students have graciously agreed to share their work, some of which is included in the following pages. For this generosity, we particularly want to thank Grace Judge, Tom O'Mealia, Tanika Raychaudhuri, Karinne Smolenyak, and Bing Sun. They all have amply demonstrated how much students learn in college. We hope this book contributes to that learning.

People have asked us whether it is difficult—maybe even risky—to co-write with one's spouse. You can't generalize on the basis of a small n, but for us, the answer is no. It has been the opposite: this collaboration has been one of the most rewarding of our writing careers. So we are very grateful to one another. Oh, and also to Buffy the Vampire Slayer, Hershey, and ZuZu, for keeping things in perspective.

THINKING AND WRITING LIKE A POLITICAL SCIENTIST

How is writing in political science different from academic writing in general? As you've probably noticed while completing papers for various courses at your college or university, expectations for writing change from discipline to discipline—a great English paper and a great political science paper may look very different because they are written for different audiences. Sorting out the rules as you move from subject to subject can be confusing. This book is meant to help you understand what counts as good writing in political science, and why.

Political science is a broad and diverse discipline. Its areas of study range from the familiar—presidential elections, say—to the unfamiliar—ancient Athenian democracy or policy conflicts in Inner Mongolia. Political science departments tend to have a variety of subfields, such as American politics, political theory, and comparative politics, that focus on different kinds of political questions, and this introduces another set of writing challenges, because even within political science, different subfields ask students to do different types of writing.

If you are an undergraduate student in political science, this book is designed to help you understand how political scientists think, how research and writing differ across political science subfields, how to generate good strategies for writing your papers, and how to follow political science norms regarding citation and style. If you are a new graduate student, you may find the book helpful as well, particularly if you didn't major in political science in college, or didn't complete your undergraduate program in the United States.

In this chapter, we'll orient you to the world of the political scientist and provide you with some concepts that explain why writing differs from one political science subfield to the next.

Thinking Like a Political Scientist

We'd like to begin by introducing you to how political scientists approach their research—what it means to "think like a political scientist"—and to the contexts in which they do their writing. A "professional political scientist" might, for example, refer to a policy specialist working in a think tank, a governmental advisor on foreign affairs, or a professor conducting research on electoral behavior. Much of the writing professionals produce is intended for specialists, but some is also directed toward the general public. For example, an advisor at the U.S. Department of State might write policy memos that take certain kinds of advanced political knowledge for granted, while a television personality like Melissa Harris-Perry almost has to hide her academic training and expertise in order to be accessible—and entertaining—to the viewers of MSNBC. Undergraduates may occasionally be asked to write for similar (imagined) audiences, but most undergraduate writing is directed toward instructors and peers; it is used to help you develop more sophisticated ways of thinking about political questions and discuss them with other engaged and literate non-specialists.

Subfield Differences in Political Science

Depending on your college or university, you might be taking courses offered by a Department of Political Science, Department of Politics, or Department of Government; you might encounter political science content in international relations, public policy, public administration, and international studies courses. Regardless of the name of the department, political science is divided into subfields that are reflected in the profession and, more important for you, in undergraduate and graduate curricula. A common subfield division in the United States is shown in Table 1.1.

TABLE 1.1 Common Subfields of Political Science

American politics	The study of governance, other political institutions, and political behavior in the United States
Comparative politics	The study of governance, other political institutions, and political behavior outside of the United States
International relations	The study of relationship between countries, and of supranational institutions
Political theory	The study of the historical and contemporary theories, values, ideals, and concepts guiding political thinking

Let's elaborate a bit.

- **American politics** studies, as you can guess, structures, institutions, and practices in the United States. This includes legislative institutions (such as Congress), the executive (such as the presidency), and judicial institutions (such as the Supreme Court). Sometimes, the study of law and politics has its own subfield, variously called "law and politics," "public law," "law, courts, and politics," or even "law and society." The role of federalism—the relationship between the federal government and the states—is an important subject of study, as are the politics of states, cities, and other localities. American politics also studies elections and related behaviors, attitudes, and beliefs of voters, parties, and interest groups, and how political communication works. An "Americanist," as such a scholar is called, might try to answer his or her research questions through quantitative means, for example by creating a "score" or an index to categorize the positions of politicians or judges, but he or she might also interview political actors or even embed himself or herself in their workplaces to observe their daily lives (this approach is called "ethnographic"). Some Americanists also use the tools of psychology, such as experiments designed to see whether racial bias might be triggered by campaign ads.[1]

- **Comparative politics** is interested in many of the same questions as American politics but focuses on countries other than the United States. Some "comparativists" study specific areas of the world (e.g., Africa, Europe, Central Asia) or even single countries (such as China), while others might orient themselves in terms of themes, such as political economy or authoritarianism. Everyone's work is still comparative in the sense that it contributes to a general understanding of the differences electoral systems or regime types make for economic development or democratization, for example. Because they often have deep knowledge of specific areas of the world, comparativists frequently contribute to mainstream media discussions of current events, in addition to publishing academic work.

- **International relations** (IR) is often confused with comparative politics by people outside political science, but its focus is more on what happens *between* countries (and other international actors, such as the United Nations or the World Trade Organization): wars and other conflicts as well as international trade. There is overlap between these subfields: many questions of comparative political economy are also questions of international political economy, for example, and a national ethnic or religious conflict of interest to those studying comparative politics might also have international implications. An IR scholar might write an academic article that analyzes the Cuban Missile Crisis between the United States and the Soviet Union as if it were a game in which various moves can be analyzed using formal mathematical tools, but an IR scholar might also write an op-ed article for a newspaper, arguing that talking to terrorists might be a defensible policy.[2]

- **Political theory** sometimes confuses people, too, because the scholarly work in all subfields of political science involves some theorizing (we will talk about the many meanings of "theory" in Chapter 3). As a subfield, however, political theory focuses on political concepts ("what is democracy?") and political values ("what is just?") and often makes normative arguments—arguments about what *should* be true, rather than descriptions of what is true—about those concepts and values ("the ideal form of democracy is representative, rather than direct"; "a gendered division of labor within families is unjust").

There are many variants to the subfield categories we just described, and many varieties within the subfields. These don't cover everything in political science, either. For example, political methodology, or the study and further development of the methods political scientists use, is its own subfield in the profession and in graduate curricula; sometimes, undergraduate curricula also include methodology as a separate subfield. Public administration and public policy are sometimes subfields, sometimes separate departments, sometimes even a different "school" or "college" within a university.

The methods political scientists use can be grouped in different ways. A common distinction is between *empirical* and *theoretical* approaches. There are two related questions embodied in that distinction. One is, "What kind of stuff do you look at in your work?"; the other is, "What kind of information tells you whether your ideas were correct?" Let's call those the *objects of inquiry*. In empirical approaches, the answer to both questions is "events, practices, and behaviors in the world."

In theoretical approaches, the answer is "relationships between ideas." So empiricists might explore what kinds of factors correlate with democratization or how members of Congress think about their constituents, while theorists might study what the implications are if people are naturally selfish or if the power balance between two nuclear-armed states is equal.

> *Object of inquiry* = "The stuff we study."
> *The question of method* = "How we study it."

We can also map a different grouping onto the examples in the previous paragraph. Here, the question is: "How do you study it?" Let's call that *the question of method*. Different people characterize the distinction in two different ways, but for our purposes, let's lump them together: quantitative-formal versus qualitative-interpretive. A scholar interested in the correlates of democratization and another interested in modeling conflict between two rational agents might both use mathematical tools. A scholar shadowing members of Congress and another thinking about natural selfishness both interpret things (even if one is interpreting what live people say and the other what Thomas Hobbes wrote in 1651). Table 1.2 offers examples of the kinds of approaches these two dimensions might generate.

TABLE 1.2 **The Question of Method**

		Quantitative-Formal	Qualitative-Interpretive
Object of Inquiry	**Empirical**	(1) Statistics	(2) Archival Ethnography Interviews
	Theoretical	(3) Mathematical modeling	(4) Conceptual Textual

In practice, many political scientists combine these for a "mixed-methods approach." For example, they might use interviews with lobbyists to deepen a statistical analysis of lobbying in Congress, or they might use archival research as evidence that a formal model of a historical event is plausible. Political science is, in other words, a "methodologically pluralist" discipline.

Later in this chapter, we'll try to give you a sense of how these questions of method and objects of inquiry are taken up in the different types of writing professionals and students produce in political science.

Thinking Like a Writer: Understanding How Reader Expectations Affect Rhetorical Choices

Above, we outlined what it means to think like a political scientist. To make the best possible use of this book, you also need to know how to think like a writer. To us, "thinking like a writer" means learning to see the variations between different disciplines and different types of texts, and understanding a few key concepts that help explain those variations. Just as political scientists have specific ways of thinking, they also have specific ways of writing, and of reading.

Let's start with the reading. You might be surprised to learn that political scientists writing in different subfields don't just have different styles of writing, they also have different habits of reading. Those who work with quantitative data (on voting behavior, for example) may expect to see prescribed sections ("Introduction," "Methods," "Results," and "Discussion"), while political theorists would be surprised to encounter such headings. There are many other expectations that also shape a reader's understanding of what counts as good writing, including expectations about vocabulary, citation styles, and the "right" ways to convey disagreements with other writers.

We'll talk about the differences between readers and types of writing in terms of rhetorical concepts: *audience, purpose, genre,* and *credibility*. First, it might also be helpful to say what we mean by *rhetorical concept* (Table 1.3). We know that many people use the word "rhetorical" to refer to speech or writing that is either pointless or underhanded, but we're using it in its more classical sense: rhetoric is the art of persuasion, using words (whether spoken or written) or images as tools to get others to agree with us.

Rhetorical = ~~sneaky things sneaky people say to get their way~~
Rhetorical = pays attention to what is relevant and convincing to specific readers

It might not seem obvious at first that "the art of persuasion" applies to academic papers that don't lobby for some position or other, but think about it this way: when you write a paper, you need to persuade your reader (and, in the classroom context, your grader) that what you've written is worth taking seriously— that your information is correct and your arguments make sense. Aspects like those mentioned earlier, including seemingly trivial elements such as your choice of citation style, can affect a reader's impression of the quality of your data and arguments. When your work doesn't match a reader's expectations, the reader is less likely to trust you. We'll say more about this later.

TABLE 1.3 **Rhetorical Concepts**

Audience	The readers you want to persuade
Purpose	Why you are writing
Genre	The form your writing takes
Credibility	How trustworthy and authoritative your readers think you are

Audience

From a writer's perspective, the audience for any piece of writing includes all of its intended (and imagined) readers. It's often impossible to know for certain who will read your work, but a writer's beliefs about his or her readers should shape every aspect of a text. Imagine describing campaign finance laws to your physics professor. Now imagine describing them to a fifth-grader. These descriptions are likely to differ greatly in terms of length, complexity, and what we might call "register," which has to do with appropriate choice of vocabulary and syntax and is often referred to in terms of "formality." You would probably write in a much less formal style for the fifth-grader than for the physics professor, using simpler sentences, more common vocabulary words, and a more casual style overall.

Purpose

A paper is shaped by its author's purpose for writing it. A political scientist might write to provide crucial information to a diplomat heading into a negotiation, to convey new ideas to fellow research specialists, or to explain polling data to the general public. Note that purpose is always tied to audience: one writes for specific readers, and for specific reasons. When writing a paper for a class, your purpose might be "to get a good grade," but to do that you often need to write as if you have a different purpose in mind. Even if your driving objective is a good grade, you should be able to articulate a purpose for your paper that would matter to readers who won't be grading you; such a purpose may be intellectual, practical, or both.

Genre

According to John C. Bean, "genre refers to recurring types of writing identifiable by distinctive features of structure, style, document design, approach to subject matter, and other markers."[3] Examples of genres include scholarly articles, research proposals, personal essays, movie reviews, blog posts, and tweets, among many, many others. However, genre is about more than just format. Different genres serve different purposes for different audiences: scholarly articles serve the purpose of conveying research findings to fellow specialists, while movie reviews evaluate films to help moviegoers decide what they want to watch. One difficulty for students taking classes in multiple subjects is that the "same" genre—the scholarly article, for example—can change from one discipline to the next because the "same" audience (fellow specialists) looks different once we get more specific. Biologists and literary critics and political scientists may all be "research specialists," but they have very different reader expectations.

Credibility

Also known as "ethos," credibility refers to a writer's trustworthiness and authoritativeness, as perceived by his or her audience. Liars are not credible; nor are people who don't know what they're talking about. By now you're probably seeing how closely connected these rhetorical concepts are: *writing is only credible when it meets its audience's expectations for its genre, and writing that is not credible will not achieve its purpose.* Credibility must be earned, and a big part of this book is showing you the tried-and-true strategies that experienced writers typically employ to earn it.

Common Genres of Political Science Writing

Scholarly writing in political science reflects the discipline's methodological pluralism: in interpretive or (non-quantitative) theoretical work (quadrants 2 and 4 in Table 1.2), much major research still appears in monographs (books on single topics), whereas in "empirical" and "quantitative-formal" research (quadrants 1, 2, and 3), individual "Introduction, Methods, Results, Discussion" (IMRD) articles are often expected. As you may have noticed, quadrant 2 appears in both categories; this reflects the variety of research approaches encompassed in that quadrant.

At the student level, writing reflects but also transcends this methodological diversity: in addition to "academic" writing, political science courses frequently use policy memos, op-ed articles, and blogs to help students practice their analytic skills (see Fig. 1.1). Conventional academic

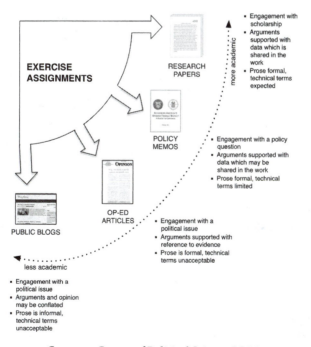

FIGURE 1.1   Common Genres of Political Science Writing.

writing in political science courses most commonly asks students to respond to, analyze, and compare theoretical and empirical claims. A common term paper across subfields is a theory/case application of some kind: "Use x event in y country to explore the theory that p causes q," or "Given that this theory predicts x, why does y keep happening in z-like contexts?" Although most political science courses do not ask students to do the kind of research that requires them to collect their own data, some do (honors and senior theses are the most common cases), and courses increasingly ask students to design research or analyze and write about pre-existing data. In some courses, "writing" also includes data visualization.

Over the past decade, as the Internet has lowered the bar to publishing, political scientists have flocked to the blogosphere and social media along with the rest of the world. In fact, the increasingly high profile of some blogs that political scientists contribute to shows that this is a genre readers come to seeking political science expertise.[4]

In addition to the types of writing described above, which emulate what "real" political scientists do, at least to some extent, quite a lot of writing in political science comes under the umbrella of "apprenticeship genres": short papers in response to readings or "thinking prompts," exam-like responses to narrowly defined questions, or "finger exercises" to practice a particular general skill, such as summarizing, paraphrasing, or reviewing literature. These are called "apprenticeship genres" because they don't appear outside the classroom setting; they are tools for communication between students and instructors and don't resemble genres that are intended for publication. In general, we regard these apprenticeship genres as valuable, but there are better and worse ways for students to approach them, which we will illustrate in Chapter 2.

What do "short" and "long" mean in political science courses? Many instructors—and students—think in numbers of pages, although varieties of new digital media formats are making the typed, double-spaced, printed, letter-sized page just one output mode among many. In this book, we will mainly speak in numbers of words; the common conversion is roughly 250 words per one typed, double-spaced page. Short assignments are generally 1,000 words or fewer, medium-length assignments usually comprise 1,000 to 2,500 words, and longer assignments, such as term papers, are more extensive. Except for theses, which can run anywhere from 6,000 to 30,000 words, a single assignment in a political science undergraduate course rarely exceeds 5,000 words.

Common Genres in Political Science Courses

Compare and Contrast

You'll see "compare and contrast" papers in pretty much all college courses. Students often think that an invitation to compare is also an invitation to play the referee—that is, that you decide which of the things you compared is right, better, or most worthwhile, and which is wrong, worst, or dumbest. Most of the time, though, that is not what the instructor wants you to do. Your goal is to focus on the differences between the things you compare and to think about the implications of those differences. *Sometimes* you will be asked to evaluate (make judgments), as you'll see, but more often not.

- **Comparing ideas.** In a political theory course, you might be asked to compare the social contract theories of Thomas Hobbes and John Locke, for example. Both theorists developed a theory of "the social contract," but their theories are very different. The assignment might ask you to spell out the differences first and then use the texts to speculate as to *why* the theories are different. Or it might ask you to draw out the political implications of the differences in these social contracts. In an American politics course focused on U.S. Constitutional interpretation, you might be asked to compare the *reasoning*, instead of the theories, behind the majority decisions in two or more Supreme Court cases.

- **Similarities and differences.** There are more than two sides to most issues, and a common assignment asks you to look for similarities and differences among *several* ideas or phenomena. This is a typical structure: "Think about our theorists/ideas/events *x*, *y*, and *z*. Build an argument that shows the similarities between any two *in contrast* to the third." In most assignments of this kind, the instructor isn't looking for "the" answer, but for your ability to think about how different ideas or events could be analyzed and classified.

- **Analysis and synthesis.** The basic "compare and contrast" paper is a form of *analysis*: breaking something—a theory, for example—into smaller parts and asking how and why those parts bear the relationships they do. In the "similarities and differences" variation above, analysis also helps you create a *synthesis*: connecting smaller components into something larger, such as categorizing Marx's and Durkheim's methods (the smaller bit) as instances of "structuralism" (the larger bit). There are lots of ways, of course, in which you can do both; often the purpose of these kinds of assignments is to see if you have understood the relevant features.

Literature Reviews
Research papers, which we'll discuss later, necessarily involve literature reviews, but often instructors assign such reviews by

themselves. Addressing the existing literature is an important way of entering "the scholarly conversation," which we discuss throughout this book. Your instructors don't usually expect you simply to report what so-and-so and such-and-such said; literature reviews are an important form of *synthesis of what is known about your topic* and also an *analysis of what the outstanding questions are.*

Research Papers

People used to think that research meant "go to the library and look up stuff." Now, people think research means "look up stuff on Google." Both are, strictly speaking, misconceptions. Sure, research can and should involve looking things up, whether in physical or digital holdings. But by "research," academics mean answering questions nobody has yet answered, or providing better answers than we already have. Such a demand for originality is a tall order in practice, and most undergraduate research, especially that done in semester- or quarter-long courses, never achieves it. It's not even the goal of most research assignments at the undergraduate level or during the first years of graduate school. But keeping the idea that "research is answering a question nobody knows the answer to, or answering a familiar question better" in view helps you understand what your instructors are trying to teach you: to think like a political scientist. Looking up stuff is simply one step in the research process. It is usually a way to find out which questions have already been answered (and which haven't), how others have (so far imperfectly, you think) answered your question, and why your readers should care about finding a better answer.

Research Proposals

An academic term, even a sixteen-week semester, is generally not enough time to complete a significant research project, especially if you are expected to learn about the substance and methods of the research during the same semester. For that reason, you won't often encounter assignments that require entire projects. But a research *proposal* is a relatively common assignment. It is also by far the most common "backstage" genre for most academics. By "backstage" genre, we mean writing that happens away from the limelight but that is nevertheless very important. Research proposals are everywhere in the business of political science: in the form of thesis proposals to gain permission to start a big project or to get an advisor, grant proposals to get financial support for a project, conference proposals to get audiences for your work in progress, or book proposals to secure a publisher.

Research proposals are also a great assignment to help even the non-professional political scientist—you—practice and demonstrate important skills. They demand an ability to engage in analysis and synthesis by efficiently characterizing existing scholarship on your topic and situating your question within it and, even more important, by showing how the research method you choose can help answer the question you are asking. They are also an important test of your argumentative writing skills: you need to convince your audience—a real or imagined funding agency, for example—why your question matters and why you are the person to pursue it. For that reason, the research proposal is the genre that probably has the broadest usefulness for your future, even if that future will take place very far from political science: as a smart college student, you will almost certainly have to make the case for projects you want to undertake, whether these are movie pitches, business proposals, or engineering projects.

Because even a research proposal that is not followed up by carrying out the proposed research is a complex and demanding assignment, you will often see it split into several components (this is called "scaffolding"). For example, one political scientist scaffolds a grant proposal project in his American politics course with two preliminary assignments: an early "idea memo," in which the student explains his or her question, what method the research would use, and why it matters, and a midterm "progress memo," in which the student reports on what has already been done, what problems have arisen, and what still needs to happen. Other courses might scaffold the proposal by having a staggered set of deadlines for different common components of a research proposal: an annotated bibliography might precede and inform a literature review, which in turn might precede and inform the description of why the proposed project matters.

A research proposal does not necessarily need to be scaffolded. Another instructor simply provides a very detailed assignment prompt that enumerates what is expected (the general theory, a testable hypothesis implied by the theory, the variables relevant for the hypothesis) and explains what those components are.

Data Analysis

Research proposal assignments are often designed to get students to understand and work with *data*. Many political science curricula offer scarce resources for methodological training, but sometimes undergraduates do have opportunities to do advanced work that involves original data collection or to work with data sets made available by others.

Not all data is quantitative, of course. But it is also important to understand that even when assignments like this involve numbers and use concepts such as "descriptive statistics," they are not primarily about math or statistics. So even if you (think you) are a person who doesn't like math, don't avoid courses that require you to use quantitative data. Sure, learning math *is* valuable, and down the line, if you continue in political science, mathematical and statistical training can come in handy. But good research assignments get you to understand the relationships between *theories and evidence*; the numbers are just a way of getting there.

Consider, for example, this assignment for a course called "Energy Politics." The instructor asks his students to produce a graph or a scatterplot that visually conveys the world's countries' electricity consumption per capita versus their GDP per capita and asks them to explain any patterns they observe. This is a writing assignment, not only because it asks the students to describe the relationship between consumption and population in words, but also because the visual element, whether a graph or a plot, is itself a way of making a claim, a piece of visual rhetoric. (See Chapter 4 for more on visual rhetoric.)

Response Papers

Response papers are a common assignment in courses at all levels. They may have several purposes, perhaps simultaneously. Sometimes, your instructors will use them in place of quizzes to make sure you've done the reading and perhaps to evaluate what you have gotten out of the readings. They may also be circulated to other students in the class, as springboards for in-class discussion, for example. Or they may be your instructor's way to prime your thinking at the beginning of a process that might culminate in a term paper. In political science courses, the key to response papers is that the instructor is usually *not* looking for your political opinions but is interested in hearing about course-related ideas and questions a reading raises for you.

Applying Theories to Cases

As we noted, none of the examples above ask you to "go outside the text." But political science is primarily and ultimately about the world, about social and political phenomena. The discipline's theories are attempts to explain *why* or *how* the world works in the ways it does, so you'll often encounter assignments in which you are asked to use either analysis or synthesis to connect ideas with facts about the world.

Many political science assignments won't ask *you* to explain phenomena in the world, but they will ask you to discuss existing explanations. Another version of this kind of assignment might ask you to identify a case or several cases in the world to which a theory applies. In an IR course, for example, the assignment might say: "Identify a war that illustrates the theory of bureaucratic stovepiping, being sure to elaborate on the features that make it such a case." In a basic political science course, the instructor or your texts have probably discussed such cases, and your task is just to recall and recognize the relevant features ("First World War!" you exclaim). In a more demanding variant, you might be asked to do the digging to find such a case: "We discussed the British military strategy in WWI as a case of the stovepipe problem. Can you offer another example of the same phenomenon? What makes it such a case?"

Advocacy Papers

In Figure 1.1, both policy memos and op-ed articles can be called advocacy papers. Sometimes these are also called "position papers," but we prefer "advocacy" as a term because you may take positions in more "purely" academic papers, that so-and-so's theory of democratization better explains post-Cold War Eastern Europe, for example. The distinction is small but important: in advocacy writing, you try to get someone ultimately to *do* something, such as getting a government agency to adopt a policy, or persuading voters to support a candidate.

Blogging in Political Science Courses

Blogging can simply be a platform for response papers, and the audience for a blog might be no wider than your classmates, or even the students in your particular course section. Some instructors use this kind of blogging to prepare or continue in-class discussions. Blogging may also be a form of "journaling," another form of low-stakes writing meant to keep you thinking about course themes or to help you develop your own bigger writing projects.

Blogs aimed at wider audiences, perhaps open to the whole world, offer a very different way of helping you practice writing.

In such cases, the blogging may still be low-stakes in the sense that any single post or comment won't play a large role in determining your grade, and instructor feedback may be limited, but of course writing that your parents—or future employers—can read raises the stakes in a different way. If your instructor makes a course blog public and promises (or threatens) to keep it live, you will want to think about your posts as carefully as you think about which party photos to share through social media for posterity.

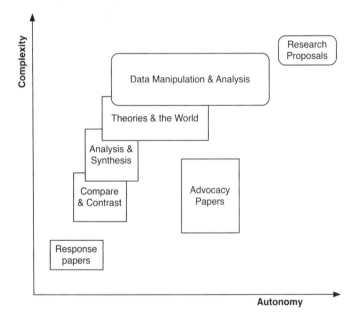

FIGURE 1.2 Assignment Types, Roughly Grouped by their Complexity and How Much Independent Creativity is Expected.

In Figure 1.2, we offer a rough mapping of how assignment types increase in difficulty. Here, we see difficulty increasing in two dimensions: in complexity and autonomy. By "autonomy," we mean the degree to which instructors expect original work and thinking from you. By "complexity," we mean an increase in the number of "moving parts" involved in the work. More complex papers require writers to manage more information, facts, concepts, and theories, and more relationships between such components. This includes "getting things right," which students sometimes think is the only thing instructors care about, but it's more than that. (That's why instructors sometimes say, "There is no one right answer, but there are many wrong answers.") If this seems daunting—or if these concepts are all Greek to you—don't worry! We will talk more about the specific mechanics of research and writing in the following chapters.

How to Use this Book

In the rest of this book, we will give you more detailed guidance on how to tackle your writing assignments. In Chapters 2, 3, 4, and 5, we'll walk through strategies for drafting and revising papers. Chapter 2 helps you figure out what your writing assignments are asking you to do, and Chapter 3 walks you through the process of writing a complete paper, with a focus on literature reviews and research papers. Chapter 4 offers guidance to those of you who find yourselves with an opportunity to write the kind of research paper for which you collect and analyze your own data, or to write a proposal for doing this kind of research. Chapter 5 provides advice for writing a few other genres that are common in political science: response papers, applications of theories to cases, advocacy papers, and blog posts. Chapter 6 focuses on details of language and style—writing clear prose and stating your claims in ways that lead readers to trust you. In Chapter 7 we'll explain how to choose appropriate sources and cite them effectively. Finally, a brief appendix offers advice about getting feedback on your writing, and another lists sources that may be useful to students who need to collect their own data or create visual representations of data.

DECODING YOUR WRITING ASSIGNMENT

In Chapter 1, we described how political scientists think, what types of papers they write, and what they see as "good writing" and why. In this chapter, we'll describe one of the central aspects of the writing process, regardless of what kind of paper you'll be writing. Because political science includes so many different kinds of genres and different rhetorical approaches, it can often be daunting to figure out exactly what your professor wants. So let's start at the beginning. For students, that's the moment you receive your writing assignment.

Decoding the Assignment Prompt

Based on what you learned in the previous chapter, you won't be surprised to hear that the first things you need to understand about your writing assignment, if you are to do a credible job of completing it, are its **audience**, **purpose**, and **genre**. Keep in mind that in the classroom context the genre may be a hybrid or invented one your instructor has created to give you an opportunity to learn a complex idea or practice a particular skill. One key question it is always helpful to ask when decoding your prompt is what this assignment is designed to help you learn. Often, your instructor will tell you this up front, as in this example of an assignment from an international relations course:

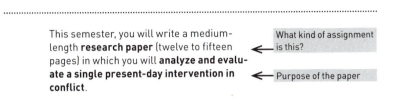

This semester, you will write a medium-length **research paper** (twelve to fifteen pages) in which you will **analyze and evaluate a single present-day intervention in conflict**.

What kind of assignment is this?

Purpose of the paper

Goals of the Assignment

This assignment is expressly designed to foster your intellectual development as a student of conflict. In particular, there are four goals you should have in mind as you develop your paper:

- Learn how to analyze and evaluate conflict in light of theory and research we study in class.
- Develop a more in-depth knowledge of a particular conflict, area, or type of intervention that appeals to you.
- Improve your research skills, including use of primary, secondary, and online resources.
- Sharpen your analytical writing style, **in preparation for future work in this program** and a potential postgraduation **job as a program analyst,** for example.
- **Prepare to write policy and research reports on conflict-related issues in potential jobs**.

Suggestions about audience, mentioned in bold

A twelve- to fifteen-page paper represents a significant time investment on your part, so it's good to be aware of the kinds of skills and knowledge you should gain from it, and to know how the instructor thinks the writing skills you're developing might transfer to the world beyond his or her course.

The genre you'd be dealing with in this assignment is stated up front: This is a research paper. An audience beyond the instructor is not specified, but given that one of the learning goals mentions preparing for work as a program analyst, it makes sense to imagine one's readers might include both academic specialists and people who need to make real-world policy decisions. The paper's central tasks, analyzing and evaluating an intervention in a conflict, also suggest its purpose: to determine how well the intervention in question is working, and perhaps thereby to come up with ideas for improving this particular intervention as well as others that might be called for in the future.

Contrast this with another research paper assignment:

....

Research Paper #1: Experimenting with Campaigns

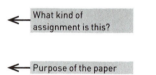

For this first assignment of the term, you are to **do an experiment**. The assignment will unfold in three parts:

1. Create two campaign artifacts for a candidate, or for an issue, that you might send out to people. These can be print advertisements or videos. The two artifacts should look like each other but should vary in some small way. For instance, you might choose to use different wording to describe the candidate, or different images, or to format the literature differently. You should have some significant theoretical justification for why you vary the presentation between the two versions. The ads should be of high quality and should look pretty realistic.

2. Send these two artifacts out to people you know, using Survey Monkey. You should use a randomization function to determine which version of the advertisement a respondent gets. Each respondent should only get one

artifact. You should have at least 50 respondents for each of the two versions of the artifact. There should be survey questions asking them about the ad and how they react to the candidate or issue.

3. Compare how the different versions of the artifact work. Then, **write up the results of the experiment and offer recommendations** for which version to use.

← Writing begins in step 3.

For the actual written paper, you should describe the experiment you did and give examples of the artifacts (showing them if they are visual, describing them if they are videos). Talk about the experiment that you set up, including assessing it for internal and external validity. Why did you use the manipulation that you used? What expectations did you have for different results? What did the results show? What do they mean?

The final paper should be between 1,500 and 2,000 words long.

Both prompts ask students to write research papers, but they are quite different—and not just because one is about international conflict and the other one is about political campaigns in the United States. The purpose of the second paper, which is stated early, is to teach you the steps in conducting an experiment. Notice how actual writing doesn't begin until step 3 in this assignment. The prior steps are ones *you*, the writer, will take to collect your own data. We will talk about how to write about data you've collected yourself in Chapter 4, where we return to assignments like this one.

Whatever the nature of your assignment, make sure you understand the terms that describe what you need to do in your paper (Table 2.1). What's the difference between "analyze" and "evaluate," anyway? They both imply that you need to think carefully about a question. As a first pass, we'd say "analyzing" something involves asking why it came to exist in its current form and how its components interact (in this case, why and how the conflict began, why the intervention was chosen, and how the intervention has been implemented). "Evaluating" something, on the other hand, means asking "how well" it meets its purpose—to what extent the intervention is reducing the level of conflict—and why it is succeeding or failing. Analyzing is a necessary step on the way to evaluating, but as we pointed out in Chapter 1, it is possible for a writing assignment to ask you to provide an analysis but refrain from offering an evaluation of whatever you have analyzed.

TABLE 2.1 **Common Verbs Used in Political Science Prompts**

Analyze	Ask why and how something works or fits together.
Evaluate	Ask how well something serves its purpose.
Argue	Take a position and support it.
Discuss	May mean "explain" or "analyze."
Critique	Evaluate strengths and weaknesses.
Explain	May mean either "recount an explanation you've learned in class" or "generate a new theory of your own for why this happened"
Propose	May mean either "generate a research plan" or "recommend a political strategy"

An online resource available from the Sweetland Center for Writing at the University of Michigan notes that "other words that may be asking for analysis are *elaborate, examine, discuss, explore, investigate,* and *determine.*"[1] The differences can be small and tricky. "Exploring" a question may allow you to approach it less rigorously than "analyzing" it requires you to do, for example, and "determine" sounds to us like it spills over into the terrain of evaluation. When in doubt, ask your instructor for more information about what he or she is expecting.

In fact, we recommend asking your instructor for more information about any aspect of an assignment that isn't clear, including the format and expected citation style.

Questions You Might Ask an Instructor About Writing Your Paper if Your Prompt Doesn't Tell You the Answers:

What is the paper designed to help you learn? You'll need to phrase this carefully, perhaps something like, "*I want to get everything I can out of this assignment and do it well, so I'm wondering if you could tell me what knowledge and skills I should be developing and demonstrating in my writing.*"

Is there an imagined audience for this paper beyond the instructor? Congress? A particular scholar? The general public?

Should more space be given to some parts of the paper than others? For example, if you are asked to analyze and evaluate something, should the analysis and evaluation be given equal time? Or is one of these more important than the other?

Could the instructor offer a definition or synonym for any verbs that seem ambiguous (such as "discuss")?

If the paper addresses any contentious political issues, should you take a position on these issues?

What would be a reasonable number of outside sources to draw on in this paper? Note the key phrase "reasonable number." If you ask what the minimum number is, you'll sound like you're trying to get away with doing as little work as possible.

Should any particular citation style be used? Should the paper include footnotes, endnotes, or in-text citations and a reference list?

How to Handle a Confusing Prompt

What if you encounter a prompt that confuses you and your instructor is difficult to contact? Consider this vague prompt (we made this one up, but we're sorry to say that you may well run across something equally murky during your college career):

> In a five- to ten-page paper, discuss Fukuyama's argument about the end of history. You may draw on outside sources to make your case, as long as you cite them properly. Your writing should be specific and concise.

Whew! So much ambiguity here. The questions that might hit you first are the practical ones: Will a five-page paper really receive as much credit as one that is twice as long? Does "discuss" mean "analyze," or does it only mean "explain"? If you *may* draw on outside sources, does that mean it will look bad if you don't, or will it look like you are wimping out on thinking for yourself if you do? How many outside sources would be appropriate? "Specific" makes it sound like the paper should include a lot of detail, so maybe ten pages would be better—but "concise" makes it sound like you should cut to the chase, so maybe five?

To get a handle on these practical questions, we recommend that you step back and think about audience, purpose, and genre and that you ask what this assignment might be designed to help you learn.

Let's start with that last one first. The one thing that is clear from this assignment is that the instructor thinks it's important for you to understand Fukuyama's argument. That might make it tempting to put all your energy into the one thing you know for sure you need to do (demonstrate your understanding of Fukuyama's text), which may lead you into the trap of merely summarizing what you've read. However, there's a phrase in the next sentence ("make your case") that should counsel you against that. "Making a case" requires taking a position on something, offering and supporting a judgment, and doing that well requires providing analysis first. So it seems that what this assignment may be designed to help you learn is (a) what Fukuyama's argument is and (b) what your own view of his argument is. We'll come back to that second part in a moment.

This paper falls into the category we've called "apprentice-ship genres." It's an exercise for students, not the type of work you're likely to see published anywhere in the real world, and it seems likely that the only audience you are expected to imagine for this paper is your instructor. Nonetheless, we recommend that you think of your instructor as a stand-in for academic political scientists in general because thinking about that broader audience can help you avoid errors such as addressing the instructor directly or recounting things that were said in class in a way that treats your paper like a personal conversation rather than a formal writing assignment.

Because this is an apprenticeship genre, its purpose is really just demonstrating to your instructor that you've learned what the assignment was designed to teach you: Fukuyama's argument and your own view of his argument. One thing that is crucial to understand, however, is that "your own view" does not mean the same thing as "your opinion." It *does* mean an argument that you generate yourself, one that you must support with the best reasoning and evidence you can muster.

"~~Your own view~~" = ~~your opinion~~
"Your own view" = your own original argument that you support with evidence

What's wrong with opinions? You may actually encounter a paper assignment that asks you to state an "opinion" on an issue. However, that word is used differently in

continued

academic political science than it is in ordinary speech, where people often use the word "opinion" to mean "preference" or "idea I like, though I cannot offer reasons for liking it" (as in, "In my opinion, blue is a prettier color than green").

In your political science paper (or class discussion), it's not acceptable to say, "That's just my opinion" and not support that opinion with reasons and evidence. Instead, you are expected to offer a careful argument. When you say, "I think X is right," imagine the instructor following up with "Why do you think that?" She is not skeptical; she only wants to know the *reasons* for your opinion. So if you see the word "opinion" (or "view") in your assignment prompt, you should mentally cross it out and insert "argument" instead.

CHECKLIST FOR DECODING AN ASSIGNMENT

Do

✓ Make sure you understand the **audience**, **purpose**, and **genre** of the assignment.

✓ Pay attention to the words used to describe the assignment, know what they mean, and know how they differ from other similar-seeming words.

✓ Know the difference between "argument" and "opinion."

✓ Ask your instructor for clarification if something is not clear.

Don't

✓ Summarize readings without analyzing them, unless your assignment prompt explicitly tells you to do that.

✓ Offer "opinions" instead of well-supported arguments.

STRATEGIES FOR LITERATURE REVIEWS AND RESEARCH PAPERS

In Chapter 2, we helped you decode your writing prompt. In this chapter, we offer strategies for the next steps in the writing process. Writing a paper—a good one, anyway—involves many steps. We'll begin with entering the scholarly conversation you'll be participating in (we'll explain what we mean by that below) and conclude with, well, writing your conclusion. Along the way, we'll address how to craft effective introductions and support your claims persuasively.

For most papers you write, you'll need to do each of the tasks below, though the need for some of these tasks may not seem obvious to you in advance if you're not writing something that is officially being called a "literature review" or a "research paper." Bear with us! We assure you that these steps really will apply in most cases. (In Chapters 4 and 5, we offer advice about writing certain types of papers that don't fit the pattern we describe here.) You may find yourself doing a couple of these tasks at the same time, or having to go back and revisit certain steps more than once along the way. Writing is a recursive (by which we mean "cyclical") process. Doing the work required in one of these steps may help you see things you missed in the earlier steps, and that's fine: that's just the way writing works. You'll want to carefully reread and revise your entire paper before you hand it in, but revision typically occurs at many other points along the way, too.

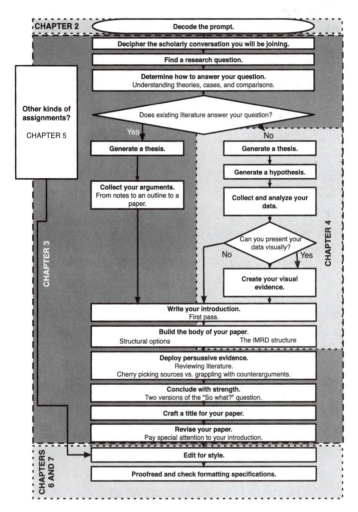

FIGURE 3.1 The Overall View of Some of the Usual Steps in the
            Writing Process and the Chapters in this Book that
            Explain Them.

Figure 3.1 illustrates our recommendation for how you should approach most papers.

Before we get started, we want to highlight two challenges that are common to political science writing. First, because of the complexity of the social world, nothing is ever conclusively proven, and competing, even conflicting, theories are common. Second, because the subject matter is political, people have strong and sometimes passionate views about it. Some of the solutions to these challenges can be met through judicious word choice, as we'll explain in Chapter 6. But the most important solution is to learn (1) to identify and operate in different registers and with different levels of analysis, and thereby (2) to develop a disposition of respectful engagement with others' ideas. This means that "thinking like a political scientist" often looks like "thinking like a person with no politics whatsoever," but this just means putting aside, not abandoning, one's politics. We'll come back to this idea at various points throughout this chapter.

Deciphering the Scholarly Conversation

In Figure 3.2, we want to stress the importance of your work's relationship to what others have done. Notice these key points:

1. In most courses, your "idea" might simply be a response to a prompt, but in more challenging assignments, you might be expected to generate the idea yourself. Anything can help you come up with it, but . . .
2. your actual paper, and therefore your plan, *must* engage existing political science scholarship.

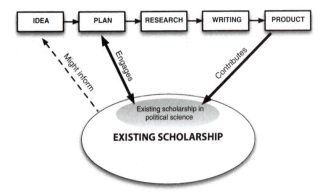

FIGURE 3.2 Your Writing Process and its Relationship to Existing
           Scholarship.

Let's return to the prompt we proposed in Chapter 2:

> In a five- to ten-page paper, discuss Fukuyama's argument
> about the end of history. You may draw on outside sources
> to make your case, as long as you cite them properly. Your
> writing should be specific and concise.

This includes an important element you will encounter in
lots of assignments: outside sources. What kinds of sources
should you use, and what are you expected to do with them?
From our perspective, the most worthwhile thing you could
do would be to use them to enter into what we would call a
"scholarly conversation" involving you, Fukuyama, and one or
two other thinkers who have responded to Fukuyama in a sus-
tained and thoughtful way. The worst thing you could do
would be to simply mine them for facts in the hope of proving
Fukuyama right or wrong. Okay, actually, that's the second-
worst thing. The *very* worst thing would be to summarize the
content of lots of other sources to take up space in your paper
because you don't know what to say.

So, back to the best thing. What exactly do we mean by "scholarly conversation"? The ideal scholarly conversation would be one in which you were able to exchange ideas with others who had thought deeply about the question at hand. The truest version of that takes place among experts who have read a lot about that question and thought about it for a long time—years rather than one semester. Yet, the point of such a conversation is that each participant contributes new insights to it, no matter how small. You can't be expected to read exhaustively enough to be sure your insights are genuinely new when writing a paper for a course, but you *can* make sure they are your own. Thus, if we received an assignment like this, we would first make sure we understood Fukuyama's argument, and we'd figure out what we thought was most interesting about it and what its strengths and weaknesses were. Then we'd look for a couple of articles by other scholars who have responded to his ideas and figure out what was most interesting there, and why, and make sure we understood their strengths and weaknesses, too. (See Chapter 7 for information on primary and secondary sources, and advice about choosing relevant articles.) At that point, we'd be ready to join the conversation.

Scholarly conversation = exchange of ideas among people who have thought deeply about a question

In a paper of this length (2,500 words maximum), you probably wouldn't have time to deal carefully with more than one or two other writers in addition to Fukuyama and yourself (Table 3.1). You might have to read more than one or two other articles before deciding which ones were worth writing about, but you shouldn't let that tempt you into believing everything you've read needs to be mentioned in your paper. Think quality, not quantity. Your instructor won't be impressed if you cite lots of texts but don't have anything interesting to say about them. Instead, he or she is more likely to see that as a sign of desperation on your part.

How to Know When You've Read Enough

Once you know what kinds of sources you're looking for and where you are likely to find them, you also need to figure out how many you need to read. This depends greatly on the type of assignment you're working on and the amount of time you have available. Sometimes, you'll be given explicit instructions ("at least two but not more than five outside sources"). When that doesn't happen, you'll have to use your own judgment. In Chapter 2, we suggested that more is not always better, and by "use your own judgment," we don't mean "read and cite as many sources as you possibly can in the time available." Instead, you should read as many as you need to in order to establish your credibility as a participant in the scholarly conversation.

In some cases—when working on a semester-long research paper or a senior thesis—you'll need to read quite a lot. But odds are, you still won't have time to read everything out there. In that situation, it makes sense to start with a few important sources and see what those writers are citing. Whom do they feel they need to respond to? Those are texts you should probably read, too. You may still find that there is a vast number of sources you'd like to read, but at the point where you find you are no longer encountering new views on the subject in question, you should stop. Your goal is to understand the parameters of the conversation, not to speak with every individual at the party.

When you're working on a shorter assignment, the same idea holds true, though you'll have to be less thorough about it. Figure out what the major positions on the topic in question are—you may have to narrow it down to the two or three most important ones—and read a representative source by an author from each perspective. Understanding and responding to each perspective well is far more important than citing large numbers of texts.

One way to make this approach efficient is to start with relatively recent research. By "recent," we mean something published in the last five to ten years. It's unlikely that anything in that work has already become the most important scholarship in your field, but the nice thing is that it will point to the most important scholarship. Start working your way back through their citations. Soon you'll be able to tell who the most significant contributors to your conversation are, not by how much those contributors themselves say, but by how much they are talked about by others.

TABLE 3.1 How Many Sources Should You Use?

2,500-word paper	Three or four total (you, Fukuyama, and one or two more)
5,000-word paper	Five or six total (you, Fukuyama, and three or four more)
10,000 words plus	The number becomes more flexible, and you may end up using dozens of sources. However, even then only three or four arguments (or types of arguments) are likely to truly matter; the most space should be allotted to these.

As we noted above, "thinking like a political scientist" often requires setting aside your own political views when writing papers. For a paper topic like the one we've been discussing, it might be tempting to decide that Fukuyama is either entirely right or entirely wrong, and to either celebrate the virtues of liberal democracy and free-market capitalism or bemoan their downsides and call for their swift demise. It's fine to take that approach when hanging out with your friends, but not when participating in a scholarly conversation. (This is an element of what we've been calling *register*.) Earlier, we noted that your job in a paper like this is to contribute an idea, no matter how small, that is new. There is nothing new, or, from your instructor's perspective, anything remotely interesting, in claiming that Fukuyama is all right or all wrong, or that capitalism is all good or all bad. Bringing a new idea to the table may sound like a challenge, and it is, but ideally it should be a rewarding one. In the rest of this chapter, we'll offer strategies to help you meet this kind of challenge, and (we hope) learn a good deal from doing so.

Finding a Research Question

Most political science writing assignments don't involve research in the sense we have defined it: answering a question to which nobody yet knows the answer. Yet, even writing a paper like the one we've been discussing about Fukuyama requires you to develop a research question of sorts. Here's one possible version: "How do the ideas of writer X and writer Y contribute new insights into Fukuyama's argument about the end of history?" That might be your working question as you get started, though you might not state that question directly in your paper. Instead, you'd state your answer, in the form of a thesis statement. We'll say more about that when we get to the section on writing introductions.

Now imagine you are asked to come up with your own research paper topic. A good instructor won't throw you into the deep end of the scholarly pool too early, but eventually you will have to come up with your own projects. For example, in the term paper assignment for his twentieth-century political thought course, Mika lists a set of themes and topic areas the course has focused on—varieties of oppression, the concept of power, political action and political ethics—and then invites the students to develop a focused paper around one of the themes, using the readings from the course.

The first step in that process is coming up with a *research question*. Developing an appropriate question is one of the most important skills a scholar can have. Most of us have interests in topic areas—the politics of inequality, or social media as a political tool, for example—but the challenge is sharpening that interest into a question that is (a) interesting and debatable and (b) something you can answer.

A good research question needs to be **interesting**, **debatable**, and **answerable**.

Your success for (a) is tested by the "So what?" question. Learn to ask yourself that question, because you can be sure your readers, whether they are your instructors, peers, or members of some other, wider audience, will ask it: "So what? Why should I care about what you are doing?" Answering the "So what?" question is tricky, though. The most common error beginning political scientists make is to be far too ambitious, which can make (b) difficult. Anybody would love to find the ideal form of democracy, solve the Middle East conflict, and have a conclusive answer to whether development causes democracy or vice versa. If you could answer those questions, the answer to the "So what?" question would be easy. You'd jump up and down and point out that they are really, really important questions. But a lot of people have tried to answer them already, and the conclusive successes have been limited. Your chances of doing that in a few weeks, in a semester, or even over a couple of semesters are slim. In fact, even professional political scientists' contributions tend to be pretty small. That's okay; that is the nature of scholarly inquiry in general. It's fine to start by thinking big, but you'll quickly need to think smaller.

Also, virtually all political science literature contributes to directly to other political science literature—those conversations we have been talking about—and only indirectly to solving real-world problems. It's not that we don't care about real-world problems, but other people have already been talking about them. So *an interesting research question is one that fits recognizably in an ongoing scholarly conversation.* You won't be able to answer the question "What explains the occurrence of ethnic conflict?" on your own, but you might be able to answer something like, "Does Varshney's theory of ethnic conflict apply outside Hindu-Muslim contexts?"

A good research question fits into an **ongoing scholarly conversation**.

So when you understand what an interesting research question is, you have also gone some way toward figuring out whether it is something *you* can answer. And, of course, sometimes you'll conclude you can't: you don't have the skills (methodological, linguistic) or resources (time, tools, research assistants, money).

That, too, is okay. We all have limitations, and knowing them is helpful. Sometimes you'll encounter fun assignments where instructors stipulate that you don't have limitations and simply ask you to propose research you might take on in a hypothetical world where you had all the time, money, and skills you could possibly need. Assignments like that reflect the importance (and difficulty) of coming up with a good question and designing the project to answer it.

Determine How to Answer Your Question: Understanding Theories, Cases, and Comparisons

The next step after coming up with a research question is to understand the relationship between theories, cases, and evidence.

TABLE 3.2   The Two Confusing Meanings of "Theory" in Political Science

In the Subfield of Political Theory	In Other Subfields
• Normative theory: how things *ought to* be • Relationships between concepts: what our concepts *mean*	• Causal mechanisms: *how* something *happens*

Theories come in different varieties, and, confusingly, the subfield of political theory understands "theory" in a way that is different from those fields of political science that don't have the word "theory" in their names (Table 3.2). In political theory, "theory" most commonly refers to a *normative* theory: an account of what something *ought to* be like. So a social contract theory, the kind you'll encounter in writing by Thomas Hobbes, John Locke, or Jean-Jacques Rousseau, says that the solution to the problems of social cooperation humans face should be understood as and arranged like a contract: treating people as equal contractors who mutually agree about something. Not all work in political theory tells its readers how the world should be, but almost invariably, it deals with theoretical, as opposed to empirical, relationships: not with what kind of strategy gets someone elected, but with the question of what it means for someone to "represent" a voter, for example.

In contrast, theory as used by most of the *other* fields of political science generally means a theory about how something happens or, to put it in fancy terms, what causal mechanisms are supposed to explain observed empirical regularities (such as incumbent advantage or democracies not fighting other democracies). Those theories look like this: "Incumbents win because their position gives them access to more financial support, and money wins elections," and "Democracies don't fight other democracies because regimes that require the potential participants in the war to consent to the war don't undertake offensive wars." Those theories might or might not be true; that they seem to fit the facts is a good sign, but not a vindication. You will learn why in your courses. Here, we are only concerned that you see that theory is a different thing from facts and data.

Understanding the parameters of a theory (the factors the theory is about) and its extension (what kinds of things it applies to) is a primary skill for a political scientist. One cannot apply a theory of voter behavior to pre-Revolutionary France because France did not have voters then; one should be leery of applying a theory of interstate conflict to civic organizations because militaries are generally institutions of states. This doesn't mean that you can't apply a theory to a new setting. In fact, often the original contribution of a scholar is to apply an explanatory theory to a new context (as we suggested in Chapter 1). But you will have to *show* that the application works. You need to argue something along these lines: "We haven't previously thought of the Jim Crow American South as a one-party authoritarian regime, but in fact applying the theory of one-party authoritarianism can help us make sense of its political development, as I will show here."[1]

Connecting a Theory to a Case

Now, if you are like most novice scholars, you won't begin with an explanatory theory, but with a specific interest in something relatively precise, a topic you formulate as a question. And the challenge you face, or may think you face, is that there are no theories that talk about your particular case. You might say, "I am interested in human trafficking in Sudan, and there is no scholarly work on that," or "I am interested in how Facebook affects political participation, but nobody has written any books about it yet." Don't despair! You *will* be able to find scholarly work on human trafficking, on Africa, and on failed states, for example, even if none of those is specifically about Sudan. (You won't find them just by Googling "human trafficking" and "theory," though.) Similarly, you'll find work, endless reams of work, on political participation, and media, and networks. What you need to do is to draw connections between these theories and your specific area. That will be your contribution; what your instructor is interested in is your explanation of how the relevant features between the cases and the theories line up.

Let's say you are working on an honors thesis about how Facebook affects political participation. (If you are, you shouldn't be surprised to hear that many, many other undergraduates are, too.) You've found the voluminous political science literatures on political communication. You should now ask yourself, how is Facebook like the media that literature mainly talks about (newspapers, radio, television)? How is it not? There is no single right answer, of course; your contribution, and the test of your learning for your instructor, is the reasoning and evidence you marshal to answer the questions. Similarly, when it comes to drawing broader, possibly generalizable conclusions—maybe about other social media that you haven't primarily focused on—you'll need to ask your questions in a new way: how is this case (Facebook as a social medium) similar to and different from those of other social media (Twitter, Pinterest, LinkedIn, etc.)? Maybe you'll realize that the private and restricted nature of Facebook makes it very different from Twitter—or maybe that factor turns out *not* to be relevant.

So here's a series of questions (Fig. 3.3) to ask every time you engage in that important back-and-forth between general theories and specific cases. (In this example, you are trying to decide whether a particular theory of political communication applies to Facebook.)

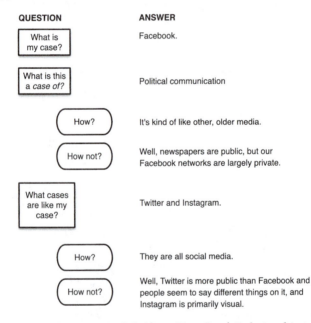

FIGURE 3.3 Questions to Ask About Your Case's Relationship to Broader Theories.

We have included the negative steps as their own questions because of their importance; they are not optional, secondary asides, but crucial for making academic arguments. That's because a failure to consider them is closely related to one of the worst scholarly vices this side of actionable misconduct: cherry picking. We'll return to that below when we talk about writing the body of your paper.

How does this advice apply in the case of the Fukuyama paper, where you are replying to a prompt? You might do the reverse: because Fukuyama is proposing a general theory about the relationship between broad ideas and historical development, you might, for example, ask if there are any specific cases—countries or historical events, for example—that don't fit the theory. Finding some doesn't mean Fukuyama is wrong, but thinking through what makes the case you found a reasonable exception or a critical argument against Fukuyama practices exactly the skills your instructors are often interested in.

Structuring Arguments: From Notes and Data to an Outline to a Paper

Almost every project, no matter how small, requires planning. So we recommend that you spend some time planning what you want to say in your writing assignment, whether it's a 250-word response paper, a term paper, or a senior thesis. In the next chapter, we'll talk about writing papers with data you collect yourself, where the writing is, in some ways, easier to plan. Fortunately, with papers where you don't need to engage in data collection, you can make your planning part of your writing. In fact, you can make your research part of your writing, too.

1. Start your outline with your goals.
   In Chapter 2, we talked about how to interpret your assignment. Once you've done that, write down the goals of the paper as elements it must contain. You might not yet know *where* in the paper they go; that's fine. It's very common for people to start their outlines with the very

generic "Intro," "Body," and "Conclusion" components, especially if they have had to learn to write five-paragraph essays. We recommend against starting your outline like that. Of course your paper will have an intro and a conclusion. And a body, too. But what you *don't* know is the structure of the body. And until you know that, you can't really know what your intro or conclusion will say. If you focus on goals instead, for a paper like our hypothetical Fukuyama example from Chapter 2, you might list goals such as "explain Fukuyama's theory," and "explain strengths and weaknesses I see in Fukuyama's theory." If you are using an outline, as we recommend, make your goals headings:

- Explain Fukuyama's theory
- Explain strengths and weaknesses of Fukuyama's theory
  - Strengths
  - Weaknesses

Outlines, Maps, Flowcharts . . . How to Plan?

Writers—students and professionals—plan their writing processes differently. Some writers create very detailed outlines, others loose and minimal ones. Some writers use concept maps or "mind maps," others flowcharts. You should do what works for you, but here is our pitch for why outlines work well in political science:

- Political science writing needs to be clearly structured and to the point. The relationship between ideas is often hierarchical; one writes about an example as a way of illustrating a point about a more general theory or phenomenon: "This theorist is a representative of communitarianism," "This regime is an example of authoritarianism." A textual outline allows you to keep those kinds of hierarchies clear.

- As our discussion of the process here shows, using an outline allows your *plan* to merge seamlessly with your actual writing: idea headings can become sub-headings and other signposts. (We'll talk about the importance of signposting in Chapter 6.)

2. Write down your initial thoughts/hunches, if you have any. You may have some ideas of what you might want to argue, or you might have different, random thoughts. Don't worry about them being random, and don't worry about whether or how they are connected. Just write them down, too, each under its own heading:
   - Explain Fukuyama's theory
   - Explain strengths and weaknesses of Fukuyama's theory
     - Strengths
     - Weaknesses
   - The Cold War as a Hegelian Dialectic
     - I think I need to check on this, but I remember from my Phil course that Hegel thought history is dialectical. Maybe I'll argue that the Cold War was some kind of antithesis to WWII. (Like, we were allies with the Soviets in WWII, then we were against them.)

3. Plan your next steps.
   What have you already read that is relevant? If this is a smaller assignment in a course, you will probably find everything on the syllabus and in your lecture notes

already. If you've taken your notes electronically, paste them into this new document, again under their own headings. Make the headings meaningfully descriptive. Not "Lecture on October 14," but "The end of the Cold War as end of history." Write a sort of "to do" list into this (now very messy-looking) outline draft.

4. Figure out what you don't need to include.

One important thing in political science is that although your instructors may often talk about the historical or sociological context for the issues they want you to focus on, they seldom want much description of these in your paper. Here, for example, you might feel tempted to describe events around the end of the Cold War or provide basic factual information ("The Berlin Wall came down in November 1989 when . . ."). In a history paper, information like that might be important, but it is unlikely to be needed in a political science paper. Political scientists often say things like, "Don't tell me what happened! Explain why it matters!"

What About Those Lecture Notes in General?

Instructors vary in their attitudes about whether and how lectures should be incorporated in papers. Obviously your paper should reflect what you've learned from the class you are taking. But should you cite lectures? *Can* you cite them? Unlike expectations regarding written sources, which we discuss in Chapter 7, instructors' views on this vary. We encourage you to ask your instructor instead of assuming anything. But two basic rules of thumb are good to keep in mind, anyway. First, remember that whenever an idea is not your own or common knowledge, you must credit its author. Second, if your instructor says, "Fukuyama believes *blah blah blah . . .*," and you've read Fukuyama, it's your responsibility to find the place where he says it and cite *that*.

5. As you read new things and come up with ideas, enter them into your outline draft *and start organizing them as you go*.

   Our colleagues who study war remind us of what all generals know: no battle plan survives contact with an enemy. The writing equivalent is that arguments don't exist until they are expressed in clear words. The nice thing about our planning process, though, is that it is "organic": it emerges as you read, think, and learn. In fact, some studies that have compared more and less successful writers have found that the more successful writers tend to rethink their argument as they go, cycle back to earlier ideas, and refine them (the less successful ones tenaciously stick to their original plan).[2] Pretty soon, you start seeing connections, so you might start grouping things you've read under the headings for the ideas you've had: your lecture notes on the end of the Cold War and the quotations from Fukuyama's work might go under your thought "the Cold War as a Hegelian dialectic."

6. Sooner or later (usually sooner!) you will realize you not only have random notes and an outline, but a partially written paper with an increasingly clear structure. And because you've grouped it around headings, which stand for the key components, you can also change it pretty easily, if and when you realize that some of the things you wanted to say don't fit

where you thought they would. You may also delete many of the things you initially included here: not all of your lecture notes are relevant, nor are all of the reading notes or quotations from Fukuyama. Your having read Hegel in a philosophy course turns out not to be relevant, so you'll drop that angle, too. That is okay. Whether you want to use a sculpting or building metaphor, it's important to realize that a good statue emerges with the removal of material and that scaffolding obscures the beauty of a building. Using a seemingly messy process like this one is still going to be more efficient than using one that goes, "I must read for three weeks, then I'll think hard, then I'll get an idea, and then I'll write it up," which often results in the "writing" part taking place late the night before the paper is due and resulting in a paper that is badly built.

Crafting Effective Introductions: First Pass

Crafting an effective introduction is challenging; the introduction may, in fact, be the most difficult part of the paper to write because it needs to accomplish many goals in a short space, as we'll explain. Thus, we suggest that you plan to write your introduction twice. The first time, you'll write one that is appropriate for what you think you're going to say in your paper; however, what you actually say is likely to end up being somewhat different from what you expect because the writing itself will lead you to refine your views along the way. As we said at the beginning of this chapter, writing is a recursive process; once you've figured out what you're *really* going to say in your paper, you'll need to revisit your introduction to make sure it prepares the reader properly for that.

We like to talk about introductions in terms of "setting up the stakes," by which we mean that you need to explain not only what question your paper will answer and how it will answer it, but also why that question matters, and to whom. This may sound straightforward, but it can be hard to achieve a balance between overstating your contribution to whatever scholarly conversation you are participating in and understating it. Sometimes, it can also be hard to judge how much knowledge you can assume on the part of your readers.

A good introduction (Fig. 3.4) must sketch out:

1. What is already known about the question you plan to answer
2. Who the major contributors to the current state of knowledge have been
3. Why it is important to know more (or see the question differently)
4. Whether you, the writer, are in a position to make a credible contribution
5. Your thesis: the primary claim you will support throughout the paper

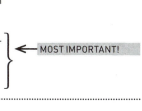

You won't assert #4 directly; rather, you'll show how you plan to go about making whatever contribution you plan to make. In any case, if any of the first three components are missing, the answer to the credibility question will be "no," no matter what approach you plan to take, and people are unlikely to read on if they don't feel they can trust you. (Instructors will read on because they have to, but they will read with a more skeptical eye than you would like.)

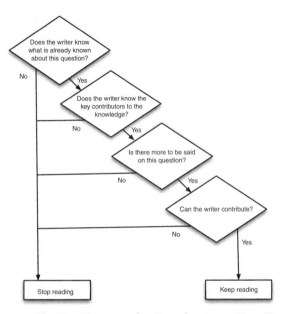

FIGURE 3.4 The Key Elements of an Introduction to Keep Your
             Reader Engaged.

Consider how political scientist David Stasavage demonstrates his knowledge of key contributors and his ability to
make a contribution in this excerpt:

Most recently, **Paul Romer has ... sug-**   ← A key contributor
gested that Europe's experience with
autonomous cities can and should be
imitated in developing countries today.
But if there are reasons to believe that city
autonomy favored European economic
development, there is also **an opposite**   ← Another key
claim. According to this view, the merchant   perspective
guilds (and in some cases craft guilds) that
controlled the governing institutions of

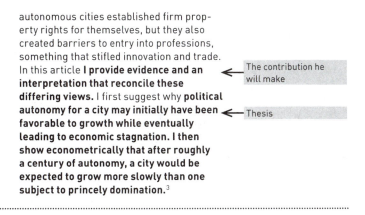

autonomous cities established firm property rights for themselves, but they also
created barriers to entry into professions,
something that stifled innovation and trade.
In this article **I provide evidence and an**   ← The contribution he will make
interpretation that reconcile these
differing views. I first suggest why **political**
autonomy for a city may initially have been   ← Thesis
favorable to growth while eventually
leading to economic stagnation. I then
show econometrically that after roughly
a century of autonomy, a city would be
expected to grow more slowly than one
subject to princely domination.[3]

Because the convention in political science is to get straight
to the point, it is virtually never wrong to have the first three
sentences of your paper state the significance of the question,
what is known about it, and how you will contribute to that
knowledge. You don't *have to* do it like that, but if you take that
kind of straightforwardness as a general suggestion, you will
do well. Here's how the beginning of an efficient introduction
to a student paper might look:

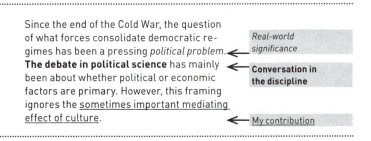

Since the end of the Cold War, the question
of what forces consolidate democratic re-
gimes has been a pressing *political problem*.   ← *Real-world significance*
The debate in political science has mainly   ← **Conversation in the discipline**
been about whether political or economic
factors are primary. However, this framing
ignores the sometimes important mediating
effect of culture.   ← My contribution

Notice that these sentences nicely capture the *two* meanings
of the "So what?" question we almost always face in political
science. The first sentence points to the *political* importance of
the question, and the following two sentences to its treatment

in the discipline. And, crucially, the writer's contribution is to the second aspect: I am not solving the political problem of democratic consolidation, the writer says, but contributing to political scientists' conversation about it.

Okay, let's say you are reasonably happy with your first three sentences. Is that the whole intro? No. What comes next, then? Ideally, some information about what approach you'll take to making your contribution, and what your thesis will be. So, the next sentence in our hypothetical introduction might be:

A COMPARISON OF TWO CASES WHERE CULTURAL VARIATIONS LED COUNTRIES IN SEEMINGLY SIMILAR POLITICAL AND ECONOMIC POSITIONS TO INSTITUTE VERY DIFFERENT REGIMES will help *illustrate one way in which the current debate has been oversimplified*, and will show that **culture, too, must be taken into account in order to better understand what leads to democratic consolidation and what impedes it**. ← My APPROACH, *contribution*, and **thesis statement**

As you can see, this introduction doesn't promise to resolve a longstanding debate, or to provide a grand answer to the debate about democracy. All it promises to do is raise a new question to add to those that are already in play, and to argue that asking that question can produce better scholarship. (In Chapter 6, we offer advice about writing style and making credible claims about what your work has to offer.) If done well, drawing attention to that question would count as an excellent contribution to the scholarly conversation.

You may be thinking that the introduction we've drafted is rather short. We agree that proportionality is important. A one-paragraph introduction might be appropriate for a three-page paper, but if you're writing a sixty-page thesis, your introduction might take five pages, or even more. Even the one-paragraph version for the three-page paper is still lacking key detail. Remember that political scientists engage other scholars *directly*: Who are the major contributors to the current state of knowledge? And what two cases will be discussed in this paper?

Introductions in political science often phrase their research questions as actual questions. Here, for example, is the very first line of an article on a comparative study of proportional representation: "Is proportional representation (PR) a desirable election rule relative to majoritarian systems?"[4] In fact, political science papers often build their questions into their titles ("Can Descriptive Representation Change Beliefs about a Stigmatized Group? Evidence from Rural India").[5] In our current example, that question might be, "To what extent, and in what ways, does culture matter for the success or failure of democratic regime consolidation?" Stating your question as a question isn't a requirement, of course, but it has the advantage of leaving no doubt about what your question actually is, so it's a good idea to make sure you could write it this way if you wanted to, even if you choose not to. If you can't, you've got a topic, not a question, and it's time to revisit the "finding a research question" section earlier in this chapter. So, a better version of the one-paragraph introduction would look like this:

To what extent, and in what ways, does cul- ⬅— Research question
ture matter for the success or failure of
democratic regime consolidation? Since the
end of the Cold War, the question of what
forces consolidate democratic regimes has
been a pressing political problem. The
debate in political science has mainly been
about whether political or economic factors

continued

are primary, with scholars such as *X* arguing the former and *Y* arguing the latter.
However, this framing ignores the sometimes important mediating effect of culture.
A comparison of two cases where cultural
variations led countries in seemingly similar ← My specific contribution to the scholarly conversation
political and economic positions to institute
very different regimes—one democratic
(Country A) and the other authoritarian
(Country B)—will help illustrate one way
in which the current debate has been
oversimplified.

A longer version of the introduction might spend a full
paragraph (or more) on each of the first four elements we listed
earlier in the chapter, though your thesis statement should
usually not be more than one or two sentences long.

In contrast, for smaller assignments, and particularly for
the kinds of papers we've been calling "apprenticeship genres,"
you might need to compress what you say about what is already known and who the major contributors to the conversation are into something more like, "What do I know about this
question, on the basis of this class, and who are the contributors I am in a position to write about?" Here's an example of a
concise introduction to the five- to ten-page paper we've been
working with on Fukuyama:

In *The End of History*, FRANCIS FUKUYAMA ARGUED ← WHAT IS ALREADY KNOWN/RELEVANT CONTRIBUTORS
THAT THE END OF THE COLD WAR REPRESENTS THE
TRIUMPH OF A LIBERAL FREE-MARKET PHILOSOPHY,
AS SUGGESTED BY G.W.F. HEGEL IN THE NINETEENTH
CENTURY. EVEN SCHOLARS WHO DISAGREE WITH
FUKUYAMA ABOUT THE DESIRABILITY OF THESE
OUTCOMES, SUCH AS RUSSELL JACOBY, PARTLY
SHARE THESE ASSUMPTIONS. HOWEVER, SAMUEL

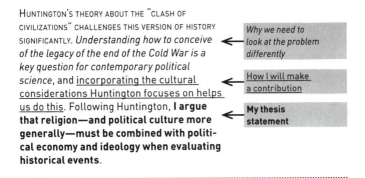

HUNTINGTON'S THEORY ABOUT THE "CLASH OF CIVILIZATIONS" CHALLENGES THIS VERSION OF HISTORY SIGNIFICANTLY. *Understanding how to conceive of the legacy of the end of the Cold War is a key question for contemporary political science*, and <u>incorporating the cultural considerations Huntington focuses on helps us do this</u>. Following Huntington, **I argue that religion—and political culture more generally—must be combined with political economy and ideology when evaluating historical events**.

← *Why we need to look at the problem differently*

← How I will make a contribution

← **My thesis statement**

The Body of Your Paper: Structural Options

"How do I know what the appropriate structure for this paper is?" you may now ask. A good question. Here's an exercise you might do at any point, not just when writing your paper.

Take a published political science article in an academic journal relevant to the course you are taking. It might be in the *American Political Science Review* (*APSR*), *Journal of Politics* (*JOP*), *World Politics, Comparative Political Studies* (*CPS*), *Political Theory*, or any such journal. Generally around 8,000 to 12,000 words, journal articles are far longer than the papers you will be writing in regular courses, but they tend to be structurally more similar than, say, book excerpts or entire books.

Now try "reverse engineering" the article by asking the following:

1. What is the author's question?
   a. Where is it stated?
2. Why does the question matter?
   a. Where is that discussed?
3. What is the author's answer to his or her question?
   a. Where is that stated?

continued

4. What approach does the author use to convince the reader the proffered answer is compelling?
   a. Is there a discussion of the approach or the methods?
   b. Where?
   c. If there isn't, why?
   d. What kind of evidence does the author use to support the argument? Textual? Empirical? Formal? (By "formal," we mean a mathematical model.)

Once you've answered those questions, you'll have a good sense of how to structure an academic paper in this particular area of scholarship. If it looks very different from what you thought your assignment asked you to do, and you are pretty confident you interpreted the prompt correctly, you are dealing with an apprenticeship genre (as we suggested you were in the case of our Fukuyama example).

In that case, we're going to assume that you will have to compare and contrast two or more arguments along the way. There are two primary ways to structure your comparisons: by writing everything you want to say about Case A (or Author A, and so on) before moving on to Case B and then Case C, or by comparing them point by point throughout the paper. Some writers find it easier to work through an entire case before moving on, while others find it easier to organize a paper according to the questions or claims under discussion. Either way is fine, but if you choose the full-case approach, you will need to remind your readers what you said about Case A while you write about Case B so they can understand the comparisons you are making. For example, with the Fukuyama paper, if you say everything you want to say about Fukuyama, and then everything you want to say about Huntington (once you decided against using Hegel, you realized Huntington was the writer you really wanted to talk about), your reader may be lost unless you draw clear connections along the way (e.g., "In contrast to Fukuyama's position, Huntington thinks the rise of anti-Western cultural forces will dampen the triumph of free-market liberalism").

TABLE 3.3   The Possibilities of Compare-and-Contrast Papers

Case-by-Case Approach	Point-by-Point Approach
Case A (Fukuyama)	**Point A**
1. Point 1	1. Fukuyama
2. Point 2	2. Huntington
3. Point 3	3. You
Case B (Huntington)	**Point B**
1. Point 1	1. Fukuyama
2. Point 2	2. Huntington
3. Point 3	3. You
Case C (You)	**Point C**
1. Point 1	1. Fukuyama
2. Point 2	2. Huntington
3. Point 3	3. You

As you can see from Table 3.3, it is important to give your own views substantial and perhaps equal time (or even more than equal time) to those of the other participants in your scholarly conversation, whichever approach you take to structuring your comparisons.

We can't stress enough the importance of giving your own ideas plenty of room. We often see an unfortunate disappearing act in student work that engages multiple sources, whether in "compare and contrast" assignments or in literature reviews, which we'll discuss below. Here's our guess as to why that happens: students lack confidence in their abilities to develop syntheses and typologies and to set their *own* agendas, or they simply haven't been asked to do so before.

Literature Reviews: Conversing with Sources and the Problem of the Disappearing Author

The function of the literature review when it occurs as a section within the kinds of research papers we discuss here and in Chapter 4 is to provide the scholarly context for your own work, to offer

> Literature reviews occur in professional political science in both books and articles. They also appear as sections within student research papers, and sometimes as complete assignments of their own.

part of the answer to the "So what?" question we talked about earlier, and thereby to set the groundwork for *your* contribution. When assigned as a freestanding paper of its own, the literature review offers you the opportunity to become deeply familiar with past research in a specific area and learn how to pinpoint and explain key similarities and differences in what others have said. (See Chapter 7 for advice on citing, summarizing, paraphrasing, and quoting directly from others' work.)

The compare-and-contrast approach to structuring the body of a paper that we described in the previous section is something you'll need to learn in order to craft a good literature review: literature reviews draw comparisons between books and articles, or groups of books and articles. They "review" (summarize and critique) the relevant literature (the body of research) that has been produced on your topic or research question so far. We should note here that **a literature review often appears early in the body of a longer writing assignment**—as the first section or chapter following the introduction in a senior thesis, for example. **However, a literature review may also serve as a subsection of an introduction** in a research proposal or in the IMRD type of research paper we discuss in Chapter 4.

Let's take the metaphor of "conversation" we have been using throughout this book. More than one kind of conversation may be relevant to your current research question—in fact, that is usually the case—so your job when you write a literature review is to let your reader in on the conversations that matter most, and to explain their key contributors and features. For example, a literature review focused on Fukuyama might include a section on theories of history, a section on interpretations of the end of the Cold War, and a section on democratization theory, among others. Below, we offer suggestions for focusing and organizing your literature review.

While we find "the scholarly conversation" to be a useful metaphor, it can also lead to the problem of the disappearing author if you don't understand the role *you* are expected to play when you participate in it. Your literature review should *not* be a report of a series of conversations you had with different authors. The conversation should be a meeting *you* are running, and to which you have invited other experts, who have roles *you* have assigned them.

Contrast the two examples in Table 3.4.

TABLE 3.4 **Contrasting Literature Reviews, Taking the Conversation Metaphor a Bit too Seriously**

Confusing List of Ideas	Clear Agenda
"First I talked to Hobbes. He said all kinds of things; here they are . . . Then I talked to this guy Locke. He had the following to say . . ."	"Mr. Hobbes, Mr. Locke, thank you for coming. I have invited you here because you are both canonical experts on the social contract. I know the disagreements among you are significant, and we may want to take note of them as we talk, but the key for the purposes of *this* conversation are the key similarities in your thinking . . ."

You obviously wouldn't take the conversation metaphor so literally as to write like either of these. But notice how the examples in Table 3.5 have exactly the same logic as the ones in Table 3.4.

TABLE 3.5   Contrasting Literature Reviews in the Way You Actually Might Write Them

Confusing List of Ideas	Clear Agenda
"Having discussed Hobbes, I will move on to Locke . . ."	"The British social contract theorists, Hobbes and Locke, are marked by their commitment to . . ."

In the examples on the left, you have no agenda of your own; you are just reporting what Hobbes and Locke said. In the versions on the right, you set the agenda by saying the social contract is the common feature (maybe your paper is about conceptions of legitimate rule, and you are contrasting the social contract to divine rule, for example). The problem with the former is that it becomes very difficult for the reader to follow the overall point of the paper because it's unclear what **you** think about the authors you're talking about. That's what we mean by the disappearing author.

The same principle applies not just to authors, but also to texts: you shouldn't structure your literature review around discrete books or articles. A good literature review will not say "Lupia's 2002 article adds to his 1994 piece the following . . .," except in the very unlikely paper in which your purpose is simply to provide a history of Lupia's work. But if you are reading Lupia, it may be because your paper is about voter competence, and a well-structured literature review in such a paper is much more likely to say something like "Against the popular and scholarly lament about the uninformed voter, recent scholarship on voter cognition (Lupia 1994, 2002; McCubbins, 2000) argues that . . ."

Here is a rule of thumb, then: imagine you are using subheadings in your literature review (a good practice in general). **If the subheadings consist primarily of proper names, you are doing something wrong:**

I. Hobbes
II. Locke

Whereas if your subheadings mention concepts, ideas, or theories, you are likely doing something right:

    I. Social Contract Theories
   II. Divine Right Theories

or

    I. Theories of Voter Ignorance
   II. Voter Cue Taking and Other Cognitive Shortcuts

> **Organize your literature review according to ideas, concepts, theories**—categories, in other words—rather than according to authors.

Like all rules of thumb, there are exceptions to this one: there are both individual authors and individual texts that may merit their very own section in your literature review. As you learn more political science, you will become familiar with the big names and the canonical texts. But the general principle is this: despite its name, a literature review is not "a report of all the things I read"; instead, it is "a discussion of how what I read matters to the question I ask and answer in my paper."

Deploying Persuasive Evidence: Cherry Picking Sources Versus Grappling with Counterarguments

Let's say you think Fukuyama got it right about the superiority of free-market liberalism, and you want to cite other political scientists who agree. In other words, you make a claim (good!) and offer evidence in its favor (maybe good). Will your reader be persuaded? Probably not—maybe *more* political scientists disagree with Fukuyama, especially now that we have some historical perspective to the end of the Cold War. If you only offer evidence that supports your position, without offering evidence that might contradict it, you have simply cherry picked your sources, and by doing that, you have *weakened* your argument because you have made your readers, at least the moderately sophisticated ones, skeptical of your skills, and maybe even your ethics. In other words, cherry picking destroys your work's credibility.

We know from experience with many students that you might find this idea crazy, but we assure you that your paper will be stronger if you take evidence and arguments against your claim into account. Think of counterarguments and contradictory evidence as a test of the strength of your claims. If you can say, "I have considered the strongest possible counterargument (or the strongest possible evidence) against my thesis, and have offered reasons why it doesn't invalidate my claims," you have made a really strong case for your ideas. After all, the political world is complicated, and almost invariably some scholars will disagree with your position, with good reasons. If you have imagined those opponents in your work already, you've gone a long way toward outmaneuvering them. For that reason, it is really important to try to imagine *the strongest* position against your argument. Picking a "straw man"—a stylized caricature of an opposing position, a position few reasonable people will actually endorse—will not help you. "Sure," your smart, real-life opponent will say, "I agree you have shown *that* challenge to your argument fails. But now consider *this!*" And then he or she will proceed to tear your argument to shreds.

"But," you might say, "I only have a week; how on earth can I know I've considered the key sources/opinions/evidence?"

Good point, especially if this is among your first ventures into a new literature. How will you know how the majority of political scientists have reacted to Fukuyama's argument? Here are a few considerations:

- *There is seldom a consensus among academics on anything.* So if things seem *too* straightforward, they probably are. Dig deeper.
- *Dig in places that best represent the discipline's mainstream.* Start your canvass of the literature with the highest-visibility journals: the *American Political Science Review* for political science generally, *World Politics* for international relations, *Political Theory* for political theory, and so on. Of course, the fact that those journals are the "flagships" also makes them controversial: don't take them as the *only* position, or the obvious truth, but simply as a shortcut to the discipline's mainstream.
- *If you are interested in individual scholars' representativeness, you can use similar shortcuts.* Citation indices tell you how much a scholar or a particular work has been cited. Again, remember that getting cited a lot doesn't mean someone is right; it simply means he or she has been influential. You may do your own informal citation index from your own course readings, too: what are the names you see a lot in the books and articles you read in this class, the one that assigned the Fukuyama paper? That can be very helpful because it will reflect your instructor's specific focuses and interests.

Writing convincingly about different positions also has a stylistic component, as we will show in Chapter 6, when we discuss the apparent paradox that weak claims make for a strong argument (and vice versa).

The No Idiot Principle

Academia can sometimes seem like a nasty place full of negativity. Many discussions are spent on how and why so-and-so is wrong about something. Professors seem to snipe at one another in articles, book reviews, and faculty meetings. Graduate students sometimes seem to adopt this negativity, too, focusing on how ideas are outdated, incomplete, even flat-out stupid. They might even imply this about the faculty they are teaching for. And you? You are invited to join the sniping, it seems! When a writing assignment calls for a "critique," you may reasonably think you are supposed to bash whomever you are engaging.

In papers that ask you to compare and contrast others' work, it may seem you should be choosing a winner and bashing the loser, or at least keeping score, but you shouldn't. Don't let the negativity get to you or to your writing. Instead, adopt what we call the No Idiot Principle. Assume that if something was put on your syllabus or assigned to you to write about, it is worth taking seriously. If you disagree with it, try to understand why. If it seems stupid, entertain the possibility that *you* are not getting it. John Rawls, one of the most important political philosophers in the twentieth century, approached his reading like that. Here's what he said about the texts he assigned in his courses at Harvard:

> I always took for granted that the writers we were studying were much smarter than I was. If they were not, why was I wasting my time and the students' time by studying them? If I saw a mistake in their arguments, I supposed those writers saw it too and must have dealt with it. But where?

I looked for their way out, not mine. Sometimes
their way out was historical; in their day the
question need not be raised, or wouldn't arise
and so couldn't then be fruitfully discussed.
Or there was a part of the text I had overlooked,
or had not read. I assumed there were never plain
mistakes, not ones that mattered, anyway.[6]

Rawls didn't, of course, agree with everything he read;
nor should you. And it is perfectly possible that there are
problems, maybe even mistakes, in the material you are
engaging. But if you begin by *not* assuming that, you will
be more careful, your rhetoric will be more respectful, and
the mistakes you find and report will be all the more compelling. Political science instructors' filing cabinets and
hard drives are full of boring, snarky, weak-to-mediocre
papers that argue that Marx was wrong because he didn't
understand human nature, Carl Schmitt was evil because
he joined the Nazi party, and the Correlates of War project
was naive in its use of very crude statistics. No instructor
wants to read another one of those. You can avoid writing
poor work by following the No Idiot Principle.

Concluding with Strength

A Common Sequence of the Components of a Conclusion:

1. Scholarly "so what"
2. Political "so what"
3. Acknowledgment of limitations of your paper

continued

4. Thoughts about future research that could address these limitations

and, finally,

5. A final claim about why continuing to pursue this approach is valuable.

Two Versions of the "So What?" Question

Once you've answered your research question and developed your argument, you may be wondering what's left to write. Didn't you already say everything you wanted to say? Too often, students conclude their papers by summarizing the previous pages, making sweeping claims, or simply stopping. But conclusions provide a useful opportunity to think about the relationship between the two types of "So what?" questions we described above—that is, the reason why the paper has been a meaningful engagement with relevant literature (the scholarly "so what"), and how it matters and does not matter politically (the political "so what"). Note that a conclusion *must* address the scholarly "so what," while addressing the political "so what" is often optional.

In the example we used for the section on introductions, the scholarly "so what" would have to do with weakness in past research and suggestions for future research, given what we've just learned about the importance of culture. The political "so what" might include thoughts about U.S. foreign policy taking culture into account when promoting the establishment of democratic regimes (Table 3.6).

Limitations and Future Directions

Another element a conclusion needs is some reflection on the limitations of your argument and/or your findings—the kinds of things you weren't able to address because you didn't

TABLE 3.6 The Two Versions of the "So What?" Question

Scholarly	Political
Contribution to political science literature: "Having identified the lack of attention to culture in the democratization literature, I showed that culture matters under some conditions."	Contribution to politics: "One implication of the argument presented in this paper is that foreign policy that remains inattentive to culture risks being ineffective."

have infinite time, money, and skill, for example, or facts or ideas you didn't take into account because you didn't realize they might matter until after the rest of your paper was finished. As we note above in the section on cherry picking and explain further in Chapter 6, showing that you're aware of your own fallibility actually makes you look smarter and more credible than trying to persuade readers your paper is perfect.

For the example we've been discussing here, one obvious limitation is the fact that we only looked at two countries. Maybe these countries were both so unusual in some way that what we learned from studying them won't apply elsewhere. (Even if you strongly believe that what you learned will apply elsewhere, it's wise to admit the possibility that it won't.)

We should note that even in longer papers, conclusions tend to be short. If you've written a sixty-page thesis, your conclusion may need to be longer than a paragraph, but it probably doesn't need to be more than two pages. Introductions need to do the difficult work of preparing a reader to understand what will follow; in contrast, by the time you reach the conclusion, the rest of the paper should already have paved the way for your final thoughts. If you've written the paper well, your conclusions shouldn't need lengthy explanations. (If you find yourself needing to provide a lot of new evidence for what you say in your conclusion, that probably means part of what you're writing there should have appeared earlier in your paper.)

Before you hand your paper in, we recommend reading your introduction and your conclusion on their own to make sure (1) you've shown the reader what you promised to show and (2) you've explained why it matters. Of course, we recommend reading the entire paper through more than once, and we provide advice on revising your complete paper later in the chapter, but it can be useful to compare these two sections on their own to make sure they line up in a way that is rhetorically effective.

The First Shall Come Last: Crafting Titles

One of the most commonly forgotten elements of student papers is the title. Many student papers either have no title or, worse, bear such woeful titles as "Short paper #1" or "Prompt 7." Titles matter. They are the author's first rhetorical move in attracting and beginning to persuade a reader.

We recommend titling your paper once you know what you have written. And unless your teacher offers specific instructions, we also recommend practicing the most common academic title structure: something pithy followed by a colon and something straightforwardly descriptive, like the subheading for this section. This is sometimes awkward, even jarring, but it is common because you want the title *both* to attract your reader (something pithy!) and describe what your paper is about (usually something not so pithy). Mika is still proud of the title of his first-ever political science term paper from his sophomore year in college: "Shit Happens: Michel Foucault's Theory of Modernity." If you know Foucault's theory, you'll know it is a perfectly appropriate title. We recommend you get to know your instructor's taste before becoming too adventurous, though.

Just When You Thought You were Finished: Strategies for Revision

Once you've completed a full draft of your paper, the best thing you can do to ensure that it does everything you want it to is revise it. Here are our top five tips for revision:

1. Put the paper away for a couple of days before you reread it. Giving yourself time to do this requires starting well in advance, but it can be hard to see the problems in something you just wrote five minutes ago. (After all, it made sense to you five minutes ago!)
2. Get somebody else to read the paper and give you feedback on it. See Appendix A for more advice about this.
3. Try reverse outlining your paper: write down the main idea of each paragraph in order. Then read your outline. Have you organized your paper as logically as possible?
4. Go through the paper and highlight each of your claims. Have you supported all of them?
5. Pay particular attention to your introduction. As we noted above, in political science, the introduction states the argument of the paper, and you won't know the true shape of your argument until you have argued it. Thus, once you've worked your way through the body of your paper, it's a good idea to thoroughly revise your intro. Are the stakes you started with still the right ones? Does your paper make the contribution you expected it to, or did it turn out to be more specific or complex than that?

    Some of the claims you made in the first draft of your introduction might need to be weakened, or strengthened, or entirely redirected, based on what you learned as you worked out the details of your argument. That's the beauty of the scholarly conversation: participating in it gives you new (and, we hope, better) ideas.

CHECKLIST FOR WRITING EFFECTIVE PAPERS

Do

✓ Situate your paper in the relevant political science **conversation**.

✓ Understand or develop the **question** to which your paper provides an answer.

✓ Develop an effective **plan while collecting your notes into an argument**.

✓ Understand **what your case is a case of**.

✓ Write **recursively**: draft your introduction twice, revisit your ideas, be flexible about your structure.

✓ Structure your **engagement of literature around your ideas and agenda** (concepts, not proper names).

✓ Address the **two versions of the "So what?" question** in your conclusion.

✓ Craft a **title**.

✓ **Revise** your paper thoroughly.

Don't

✓ **Cherry pick** your sources and evidence.

✓ Turn your opponents and foils into **straw men**.

✓ **Let your own ideas and agenda** disappear when discussing other writers' work.

STRATEGIES FOR DATA-DRIVEN RESEARCH PROPOSALS AND IMRD PAPERS

This chapter focuses on writing about data you collect yourself. We will focus on writing about data in a way that inspires confidence in your reader. Our focus, in other words, will be *rhetorical*. As we have said before, we don't mean "rhetorical" in the negative sense it sometimes has, as something sneaky and underhanded. Instead, it refers simply to the consideration of the expected forms of expression and what makes them credible and persuasive to their target audiences.

You may notice that we'll talk about the visual presentation of your data before the writing of the actual paper. There are two reasons for this. First, scholars often play with figures to explore their own data, to see what the data tells them, or whether they can see the patterns and relationships they hypothesized. Second, figures can be the rhetorically most effective way of making an argument. This may be seem to be an unusual way of going about writing, but remember that political scientists frequently first turn to the figures in their reading, especially if the data used is quantitative.

Aside from the issues of collecting and presenting data and presenting your work in certain predefined sections, all of which we'll describe here, the process of writing research proposals and IMRD papers is largely similar to that of writing the kinds of papers we described in Chapter 3. Thus, we encourage you to refer back to Figure 3.1, as well as to various sections of Chapter 3, such as the section on finding a research question, as needed.

Types of Data

Data is the plural of *datum*, a Latin word that means "given." Anybody who has collected data knows it is anything but. One of the main reasons undergraduate courses in political science don't usually teach data collection is that it involves many complicated technical, methodological, and ethical issues your instructors simply can't cover during a single academic term. You are lucky if you have had the opportunity to take a course on research methods; if you haven't, and you want to do this kind of work, we recommend that you consult a social science or political science research guidebook. (See Appendix B for a short list of books you might find useful.)

We don't have the space to explain all the ins and outs of data collection and analysis here. However, we will try to familiarize you with one of the most common formats writers use to convey their findings—the Introduction, Methods, Results, Discussion (IMRD) paper—and how to write the kind of research proposal required to explain and justify data collection plans you might or might not actually carry out. Along the way, we will offer some tips for understanding work by others—in other words, how to think like a political scientist when it comes to data—but our focus in doing so will be on maximizing the clarity and credibility of your own thinking and writing.

In this book, we treat "data" as a singular because it has become an acceptable convention to do so. However, you may encounter sticklers who insist on treating "data" as

continued

plural: "the data show . . ." as opposed to "the data shows . . ." If you do, and they are your instructors, respect their preferences. And if you're not sure what is expected, choose the more conservative option and say "the data show," even if that sounds funny to you.

One distinction between types of data is between quantitative and qualitative varieties, where "quantitative" refers to numerical data that can be analyzed with statistics, and "qualitative" refers to data that must be interpreted and categorized. The number of votes cast for each candidate during an election, for example, is quantitative; the content of media coverage of that election is qualitative.

A survey may collect both quantitative and qualitative data: numerical rankings from all respondents as to whether they "strongly agree" or "strongly disagree" with some statements (quantitative), or their thoughts, in their own words, as responses to open-ended questions (qualitative).

- Quantitative survey question: "I enjoy writing political science papers." (Rank on scale of 1–7, where 1 means "strongly disagree" and 7 "strongly agree.")
- Qualitative survey question: "Please tell us which types of political science papers you enjoy writing most and why."

Qualitative data may also become quantitative: media coverage might be *coded*, by humans or even by computers, so that it can be treated quantitatively. For example, a scholar might develop a coding scheme for language students use when they talk about what they like about writing political science papers, and then count how often certain words or phrases appear in order to compare their frequency across different subfields (Table 4.1).

TABLE 4.1   A Sample of Coded Qualitative Data

	American Politics	Comparative Politics	International Relations	Political Theory
"Close reading of texts"	0	9	1	17
"Analyzing survey data"	18	3	2	0
"Comparing and contrasting cases"	7	16	14	4

These examples hint at the virtues and shortcomings of different kinds of data. Qualitative data is rich and captures the nuances and the complexity of the world. But for that very reason, it is difficult to collect and analyze in large quantities, and therefore it can be hard to make generalizable inferences on the basis of such data. Where broad, generalizable claims are the goal, scholars often turn to quantitative data. It may be difficult to have an open-ended conversation with more than five or ten people, while it is possible to survey a thousand or more with a form that produces quantitative data. Numbers also have the virtue that they can be treated with mathematical tools: a scholar can analyze the variance of a phenomenon, test whether seeming differences are statistically significant, see whether one phenomenon correlates with another, and so on. A loss of the richness and nuance of qualitative data is the price one pays for the gains quantitative data brings.

Of course, it is important to remember that something being countable doesn't by itself mean that it permits a generalizable inference. (If 90% of the Republicans living in one small college town were in favor of raising taxes to support the local library, for example, that would not imply that Republicans everywhere would be likely to embrace similar tax choices.)

Because qualitative data and quantitative data offer different benefits, many researchers use mixed methods approaches and collect some of each. For example, to study the environmental movement, you might collect statistical data on the membership and finances of environmental groups, and do a case study on two or three such groups to get a better sense of what they actually do with their members and their money.

Discussing Data Collected by Others

In this section we'll suggest a couple of questions you might ask when evaluating data collected by others and when planning data collection of your own. (See Chapter 7 for advice about choosing credible sources of research and data.) The answers to these questions form part of the scholarly conversation that takes place in this kind of writing; you'll be expected to talk about these issues with respect to others' research, and to present good reasons for the choices you make in your own work as well. For now, we'll speak in terms of research conducted by others, and we'll use a survey as our example to illustrate the kinds of questions we have in mind.

1. How did the researchers frame the research question?

2. What is being compared to what, and why?

How Did They Frame the Research Question?

What we're really asking here is whether you can detect any obvious *bias* or *ambiguity* that may have interfered with the quality of the data the researchers collected.

1a. Bias
You might think that bias, at least, would be easy to avoid, but even researchers who want to be objective sometimes unintentionally build their own assumptions into their questions. Consider the difference between these two options:

- What makes urban U.S. voters more likely than rural voters to support increases in the minimum wage?
- Is there any difference between urban and rural U.S. voters in the probability that they will support increases in the minimum wage?

As you can see, unless a researcher has already collected data that indicate that the answer to the second question is yes, the first question is jumping the gun. It's based on what the researcher thinks he or she knows about urban versus rural residents, which may not be based on evidence. If you encounter a question like the first version ("What makes x more likely than y?"), you should first ask how someone knows that x is in fact more likely than y. If the researcher doesn't present a good answer to that question, the research may be questionable.

1b. Ambiguity
If you want to quickly see for yourself how easy it is to build ambiguity into a research question, try designing and piloting your own short survey. Pilot tests usually reveal that at least a few of a survey's questions don't make sense to their intended respondents. For example:

Confusing survey question: "Would you say the climate in Congress is improving, degenerating, or staying about the same?"

This question is trying to get at something about civility and cooperation. The researcher wants to know if respondents think members of Congress cooperate less often than they used to, which, in the writer's mind, represents degeneration. The questions sounds objective, but there are many ways respondents might interpret it:

- "Do you think members of Congress are cooperating less than they used to?" Some respondents may think cooperation is a bad thing; they may believe that their party should dig in its heels and make no concessions. Thus, even among respondents who understand what the writer meant by "climate," it is unclear what they will count as "improving" and what as "degenerating."
- "Do you think Congress is paying attention to the wrong things these days?" Because "climate" is not defined in the question above, some respondents may interpret it to mean "the political climate," or "the issues that currently seem to get attention."
- "Do you think the heating and cooling system in the rooms in which Congress meets needs an upgrade?" Some respondents may associate "climate" only with the weather. Thus, they may default to thinking about temperatures in the absence of more information.

You can see how important it is to choose *unambiguous terms* and *be as specific as possible*, without becoming too wordy. (See Chapter 6 for advice about avoiding wordiness and writing concisely.)

Improved survey question: "Do you think members of Congress collaborate across party lines more, less, or about the same amount as they used to?"

You should note that the last part of that question ("the same amount as they used to") may still lead to different interpretations. For some respondents, it may mean "the same amount as last year," while for others it may mean "as much as they did before the most recent Congressional election," or even "twenty-five years ago." So again, more specificity is required:

Final survey question: "Do you think members of Congress collaborate across party lines more, less, or about the same amount as they did five years ago?"

2. What is Being Compared to What, and Why?

You've probably heard that it doesn't make sense to compare apples to oranges (unless, of course, your question is about similarities and differences between apples and oranges). Unfortunately, however, it can be easy to compare apples to oranges (or lemons, or chocolate chip cookies) without actually meaning to. Remember our question above about urban versus rural voters' views on the minimum wage? What if all of the urban voters we surveyed lived in the North, and the rural voters lived in the South? Or, what if the urban voters were given a survey with slightly different wording than the voters in the South? Or, what if the urban voters were surveyed six months earlier than voters in the South? Any of these differences could matter enough to make the real source of differences in their opinions unclear: perhaps we really measured North/South differences, or the effects of word choice on voters' responses, or the effects of any number of newsworthy events that happened during those six months. To be sure we're really comparing what we mean to compare (the effect of living in urban vs. rural settings on voters' views about raising the minimum wage), we want all other factors to be as similar as possible, for both groups.

Once you've figured out what you think about others' data, you'll need to write about it in ways that are clear and credible. When you describe somebody else's research—which you are most likely to do in the introduction and literature review sections of your own paper—you need to state what the other writer's research questions were, what types of studies he or she did, what the findings were, what problems (if any) you see in his or her approach, and what questions his or her work has left unanswered.

Entering the Scholarly Conversation: Proposing Your Own Research

As you can see, all of these questions are also important when you're explaining research questions of your own. When you write a research proposal—whether it's for research you actually intend to carry out or not—you need to persuade your readers not only that your question is interesting and answerable, as you do with any kind of paper, but also that you have framed your question in a way that is clear and unbiased enough to allow you to gain a worthwhile answer, and that any comparisons you plan to make are justified.

Both a research proposal and an IMRD paper require you to decipher and participate in a scholarly conversation, much like the papers described in Chapter 3. It's just that in this case, part of the conversation often involves evaluating others' research methods, along with accurately describing your own.

The Introduction to Your Research Proposal

In terms of structure, the research proposal has two primary parts: an introduction and a methods section. The introduction functions much like the introductions we described in Chapter 3, where we said it needed to tell your readers:

1. What is already known about the question you plan to answer
2. Who the major contributors to the current state of knowledge have been
3. Why it is important to know more (or see the question differently)

and perhaps most importantly,

4. whether you, the writer, are in a position to make a credible contribution.

In other words, it needs to explain what your question is and why it matters, and it needs to give your readers a sense of the scholarly conversation you'll be entering. If your proposal includes a literature review, this is likely to appear as a subsection of your introduction, which may be called (not surprisingly) "literature review," or perhaps "background." See Chapter 3 for more on how to write a literature review.

Your introduction might also propose a possible answer that you think your research will find, and why you think so. If you think you know what you will find, you have a **hypothesis**. Here's a sample hypothesis (or pair of hypotheses) generated by a student in her research proposal for a research methods course:

1. There is an association between an individual's college major choice and his or her level of political participation.
2. Different types of college major cause different levels of political participation.

If this student were interested in whether college major choice makes a difference in political participation but had no prior ideas about what she would find, this research project could, of course, be purely exploratory. In that case, the research question might simply be "Is there a relationship between a student's major choice and his or her political participation?" with no hypotheses spelled out.

The Methods Section in Your Research Proposal

In the methods portion of your proposal, you need to explain how you will test your hypothesis or explore your question. What methods will you use, and why those? (In the case above, the student planned to collect survey data on individuals' college backgrounds and the wide range of political activities they might engage in.) Part of this question will already be answered in your introduction if you write the introduction well: if you've clearly described past research on questions that are connected to yours, you'll already have ideas and language available for building on past work.

Some of the questions this section should address:

- How will you collect your data? (Experiment? Survey? Interviews? Content analysis? Participant observation? Archival research?)
- From whom or from what sort of documents will you collect it? (What makes that the best source of data?)
- If you're planning to use data that has been collected by someone else, how will you gain access to that data? (And again, what makes that the best data to use?)
- Once you have your data, what analytical approach will you use to interpret it, and why? (Interpretation? What kind? Statistics? What method?)
- If you have a hypothesis, how will you know if it was right, wrong, or partly right and partly wrong?
- What resources will you require in order to carry your research out, in terms of money, space, or equipment? Will you require help from collaborators, or will you be able to do this on your own?

It is possible that you don't yet have answers to all of those questions. This may happen especially if you are taking an introductory course on research design before having taken methods courses. As always, we recommend asking your instructors what they would like to see in the section. We *don't* recommend making stuff up that you think sounds right but that you don't yet understand. Believe us, it won't sound right to your instructors.

Collecting and Analyzing Your Data

If you are both collecting and analyzing your own data, the length of this section is going to be inversely proportional to the effort you put into those pursuits, in comparison to other aspects of your work. That is, they are going to take longer than anything else in your project. They will determine what you can say—even in the early parts of your paper. Be prepared for this, go do it, and come back here when you have it (more or less) together.

We now bid you farewell for a good chunk of time.

Okay, you are back! Welcome! Now you need to write about your results. That sounds easy, and it is easier than designing your project or collecting and analyzing your data. But it is not trivial.

Because the results of your data analysis form the key contribution of your paper, you need to decide how to present them effectively before drafting other parts of your paper. This may or may not include a visual presentation of your results. The presentation of data in visual format is not a decoration that comes after the argument but (potentially) the most effective distillation of your evidence.

At this point, you confront the question of how to share your data with your reader. It is important to consider whether it will be more efficient to present it visually, or purely textually. Unfortunately, there is no clear principle or rule we can offer. So we suggest that you consider the ideas in the next section. If you feel, after reading everything, that what you have won't work as a visual, you'll have to rely on words, and to take great care in explaining your data that way.

Drawing Clear Pictures with Data: Practical and Ethical Dos and Don'ts for Visuals

Visual rhetoric (communicating with images) is increasingly important in writing of all kinds, but visual aids can be particularly useful for helping readers understand your data and findings. In this section, we will introduce some considerations for presenting data visually. Some of these are practical, such as deciding when data will be better understood if presented visually than if described in the body of a text, understanding how much material can be effectively presented in a single illustration, choosing between different types of visuals (e.g., when to use a table or photo or diagram or graph—and if a graph, which specific kind), and using labels, headings, and captions appropriately. Some are ethical: for example, the choice of a scale can dramatically (and unwarrantedly) increase the apparent significance of your results. In this section, we'll describe a few of the most common ways of presenting data visually.

Common Ways of Presenting Data

Three of the most common types of visual presentations of data are tables, graphs, and diagrams. We'll explain the differences and offer guidelines for choosing between them below. First, some quick definitions (Table 4.2).

TABLE 4.2 Type of Data Displays

Table	Data or text arranged in columns and rows
Graph	Data presented visually
Diagram	Relationships between concepts presented visually

Each of these methods helps organize and compress data so that readers can grasp relationships more easily.

Tables

Tables can be used for both numbers and words, and are appropriate when you need to display your data in terms of categories, or when no numbers are involved.

What you saw at the end of the previous section was a table about organizing information. You may also look back to Table 4.1, which presented quantitatively coded qualitative information in a *cross-tabulated* table.

Tables are one of the most common ways of presenting information visually in political science. They allow you to categorize any kind of information, and they also allow you to present quantitative data efficiently.

In fact, in political science, one of the most common statistical techniques is regression analysis, which uses probability calculus to evaluate the relationships between the variables a scholar is interested in. The results from a regression analysis are traditionally presented in a particular kind of table that looks like Table 4.3.

TABLE 4.3 **A Regression Table**

| | Dependent Variable: | | |
| | Fatality Categories (1–6) | | |
	(1)	(2)	(3)
Duration of conflict (months)	0.054***	0.043***	0.042***
	(0.001)	(0.001)	(0.001)
Level of hostility (1–5)		0.329***	0.324***
		(0.011)	(0.011)
Originator of conflict			−0.396***
			(0.048)
Constant	0.189***	−0.712***	−0.333***
	(0.017)	(0.034)	(0.057)
Observations	5,505	5,505	5,505
R^2	0.207	0.315	0.324
Adjusted R^2	0.207	0.315	0.323
Residual Std. Error	1.136 (df = 5503)	1.056 (df = 5502)	1.050 (df = 5501)
F Statistic	1,437.205*** (df = 1; 5503)	1,266.808*** (df = 2; 5502)	878.196*** (df = 3; 5501)
*$p < 0.05$, **$p < 0.01$, ***$p < 0.001$			

Using publicly accessible data from the Correlates of War project, which collects data on conflicts between countries, Table 4.3 explores the relationship between the fatalities a country suffers in an interstate conflict, on the one hand, and the duration and level of hostility of that conflict, as well as whether the country in question originated the conflict.[1] The fatalities are categorized into groups ranging from 1 to 6 (where 1 means fewer than 25 fatalities and 6 more than

999). Fatalities in this study are the "dependent variable." Duration, level of hostility (ranging from "no militarized action" to war), and origination are the "independent variables." The columns (1), (2), and (3) each represent a regression equation (a "model"); in the first, only the effect of the duration on fatalities is evaluated, the second tests the effect of both duration and the level of hostility, and the third adds the origination variable. The asterisks indicate that the relationships are statistically significant (so not a result of chance). For example, the number immediately to the right of the "Duration of conflict" says that for every one month that the conflict lasts, the fatality category increases by 5.4%. This may seem very little, and lead you to ask more questions: perhaps it matters more what the level of hostility is. And, indeed, that seems to be the case: for each increase of one in the "Level of hostility" variable, the fatality category increase is 33%. We leave as a nerdy homework exercise for you to figure out why the old saying "Offense is the best defense" also appears supported by the table.

There may still be much that seems mysterious, even bewildering, to you about that table. Don't worry. You *are* likely to see tables like this at some point in your political science career, and you may even produce your own. ("How the heck do I do that?" you might ask. The answer: with dedicated statistical software. Although the table above involves complicated mathematics, it is the result of a few basic commands from the user.)

Tables make it easy to look up information, as long as they are limited in terms of size and the number of categories they display. Imagine, for a moment, a two-column, fifty-row census table listing the population of each state. It wouldn't be hard to use this to locate the population of your specific state of interest (although it would take up quite a bit of space in your paper). Now imagine a ten-column, fifty-row census table listing the population of each state broken down by different gender and ethnic groups. Or imagine a regression table with dozens of variables and multiple models. This begins to be overwhelming and unwieldy; you might include it in an appendix if it's important to have this information available for your reader, but you would want to carve out smaller sets of this data to present in more manageable tables in the body of your paper. (See the box entitled "The problem of too much information" at the end of this chapter.)

Graphs

> Graphs are useful for condensing large amounts of data and displaying relationships between data.

Think about our made-up survey data in Table 4.2 above. The table includes just enough categories that getting a sense of what the numbers tell us gets a bit confusing. So let's present the same data in a graph (Fig. 4.1):

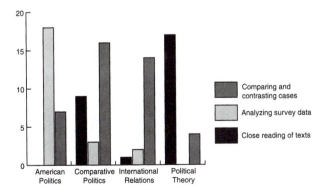

FIGURE 4.1 An incomplete Bar Graph.

Now it's much easier to get a quick sense of what the numbers tell us: there are significant differences between the subfields of political science. (Remember that we made up this data. You can't *really* tell anything about subfield differences on the basis of this.) We might say, in fact, that here the data has become *information*. That is your goal as a writer: to transform data into information.

Graphs are also often called charts. Microsoft Excel, for example, which is likely the tool you'll first use for your graphs, calls them charts. These words used to mean different things, but they are now widely considered synonymous. You'll also run into the concept of "plot," which also has pretty much the same meaning.

The graph in Figure 4.1 is a bar graph. Excel calls it a column chart. For Excel, it would become a bar chart if we flipped the axes; we might also call it a column graph. Why would we flip the axes? In this case, we wouldn't. But if political science had, say, fifteen subfields, we'd run out of horizontal real estate on the page, and making the axis with more categories vertical might solve the problem. And that's the key principle in choosing what kind of graph to use:

Use a visual that presents your data as clearly and efficiently as possible, with all the necessary information and without unnecessary distractions.

How well does Figure 4.1 satisfy our principle? Well, it looks OK. It has a *legend* that explains what the different bars represent.

Its *x-axis* tells you without too much thinking what we are looking at: the subfields of political science. But would you have
known this without reading this book or being a political science
student? And *what is on the y-axis*? Numbers, but numbers of
what? *The graph is missing some significant information!*

Good graphs are freestanding, even when embedded in
texts: they tell their viewers everything they need to know to
make sense of the data. Our graph is missing information
about what the data is all about. We told you at the beginning
of this chapter what this kind of data might be about: coded
responses regarding how students describe the writing assignments in their political science courses. We could put that information into a chart title or into its caption. Let's put it in a
title (Fig. 4.2):

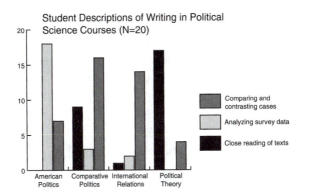

FIGURE 4.2 A Bar Graph with More Information About the Data.

There! Much better. Notice that we also added the number
of people whose responses this data represents.

Let's look at different kinds of graphs. There are many more
types than we have room to discuss, but here are some basic
ones.

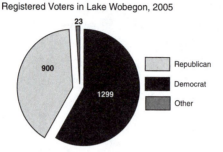

FIGURE 4.3 A Pie Chart.

Pie charts are a common way of representing the relationships between a limited number of categories, such as the number of registered voters in the imaginary Lake Wobegon in Figure 4.3. Notice that we have included the actual numbers, too, to provide more information. We could have used percentages instead, but the pie chart already gives you a rough sense of the proportions, which is what percentages are all about. Although very popular, pie charts are not particularly efficient at providing information when your data gets more complex. For example, if you wanted to show the changes in partisanship in Lake Wobegon over time, you'd have to use many pie charts, and you still would not easily be able to see the main issue.

To express changes (over time, or otherwise), you would be better off with a *line* or *area graph* (Figure 4.4).

The most prominent feature in Figure 4.4 is the sad story about the declining number of voters in Lake Wobegon. Perhaps the population of the town is declining (everyone is moving to Tucson!), or maybe fewer people are registering to vote. We can't tell on the basis of this data alone, but it is potentially significant—if, for example, we are interested in the political "weight" of Lake Wobegon in Minnesota politics. But if

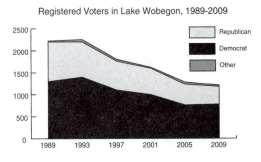

FIGURE 4.4 An Area Chart.

we are simply interested in the power balance in the politics of Lake Wobegon, we would be better served with a graph that ignores the change in numbers and "normalizes" the data (always making the proportions add up to 100 percent). That's Figure 4.5.

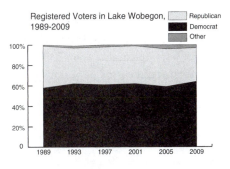

FIGURE 4.5 A Normalized Area Chart.

As we said, Microsoft Excel is the tool you are most likely to know already and therefore use to create your own figures, especially as it is quite easy to use. Google's Sheets application, like Apple's Numbers, is similar. All of them are also very limited, and most people doing serious political science research abandon these basic tools quickly, both for tables and graphs. There are many different kinds of tools available for sophisticated data analysis and graphing. Political scientists primarily use the statistical software packages SPSS, Stata, and R. (Of these, R is free and arguably most powerful, but also hardest to learn.)

Diagrams
Diagrams help you illustrate complex relationships succinctly. They tend to be underused because creating them requires time, technical skills—and creativity. And diagrams that are quick and easy are often unnecessary. Consider Figure 4.6.

FIGURE 4.6 An Unnecessary Diagram.

While in a textbook (like this one) even simple visuals can be helpful, in an academic paper, it would be much quicker to say, "The longer a war lasts, the higher the number of casualties." Also, because an arrow often implies causation, the diagram is potentially misleading, unless you really wanted to say the duration of war causes a higher number of casualties. Maybe a high number of casualties is both an effect *and a cause* of prolonging war.

Diagrams can nevertheless be helpful, and there is one area of political science where they are so common as to be expected. This is game theory: "games"—strategic, step-wise moves between various "players"—are often represented with game tree diagrams (Fig. 4.7).

As we pointed out in Chapter 1, political scientists often *model* complex real-world interactions as if they were games. Diagrams like this help make sense of the strategic logic and

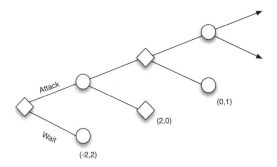

FIGURE 4.7 The Appearance of a Typical Game Tree.

the possible choices and outcomes of political events. The idea is that there are two "players," represented by the diamond and the circle. The diamond has two choices of action, attack or wait (so this is probably some kind of conflict), in response to which the circle then does something. Where the "tree" doesn't "branch," the game ends. The numbers in the parentheses are diamond's and circle's "payoffs," respectively; they represent how much each player would like or dislike that particular outcome. Learning to read and eventually to create diagrams like this is something you are likely to get out of certain kinds of political science courses.

The Ethics and Rhetoric of Visuals

Whatever tools you use—and there is nothing wrong in beginning with the basic ones, like Excel—creating graphs involves ethical and rhetorical considerations: you don't want to mislead, confuse, or distract your audience.

Consider Figure 4.8, which presents the imaginary ACT test scores for our imaginary Lake Wobegon over ten years.

Is the variation significant? There are statistical tests we can do if we really want to know, but what about what we can just see? Here is the problem: because we have chosen a bar

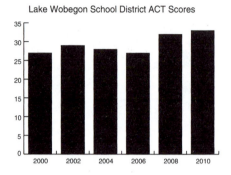

FIGURE 4.8 Bar graphs always scale from zero, which may be inappropriate in some cases.

graph, the scale on the *y*-axis begins at zero. Why shouldn't it? Well, we know that ACT scores don't vary quite that widely. It is unnecessary and, we might argue, unhelpful, and even misleading to display the data in this way. It may also be unethical if we've chosen this approach deliberately to minimize a reader's ability to perceive changes in the data. Imagine, for example, that a new school superintendent was hired in 2007. If we don't believe he has had an impact on student learning, and we don't want you to believe he has either, Figure 4.8 could be used to help make you think we are right.

However, if we change from a bar to a line graph and pick a scale for our *y*-axis that represents the reasonable variation of ACT scores—say, 15 to 35—we get a clear upward trend (Fig. 4.9). Of course, we don't know what this trend means yet. We might, in fact, need to get more data (from other school districts, or from the 1990s) to make sense of this trend.

Our point is *not* that you should pick a scale that makes your data seem meaningful. Unfortunately, this is often the default in programs like Excel: they assume you want the data

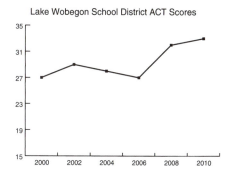

FIGURE 4.9 Line Graph with a Scale Chosen According to Reasonable Variance.

to be shown at a scale that makes differences as pronounced as possible. So, a principle to remember is this:

Choose scales and units that are appropriate for the variance of the data you want to present.

There are also rhetorical considerations that aren't quite as important as these ethical issues, but matter all the same. Another problematic behavior of software such as Excel is that the defaults offered to you are frequently the wrong ones. For example, the graphs we have produced here are all in grayscale (that is, they don't use any colors other than black, white, and the mixture of the two). This is appropriate for a book without color. Because so much of the work we do these days lives in the digital world, it is often perfectly appropriate to use color. But remember that if you will have to print your work—your term paper, say—in grayscale, be sure not to pick the default color option.

Similarly, avoid what the former political scientist and current data visualization guru Edward Tufte has called "chartjunk": don't select a three-dimensional graph because it looks cuter than a regular two-dimensional graph.[2] Except for very rare three-coordinate graphs, 3D almost never increases the visual legibility of your image—in fact, it does the opposite.

One additional ethical consideration is whether you should or can copy visuals others have produced. In a way, of course, you can: cutting and pasting images such as graphs and tables is very easy to do. And, provided you cite your sources, doing that is no different from quoting someone. In fact, pasting someone else's table or a graph into your paper *is* quoting.

Some instructors agree. But our view is that you shouldn't do this. You may use others' *data* (as long as you cite it), but lifting others' visuals and tables comes uncomfortably close to having the others do your main work—analyzing and arguing—for you.

There are excellent sources of more information about the use of visuals in social sciences if you want to improve your thinking and skills, and go beyond the off-the-shelf software like Excel. We list a few of those in Appendix B.

The IMRD Paper

As we've noted, the IMRD paper has standard sections, each of which addresses different tasks. Making sure the right information ends up in the right section may seem difficult the first few times you do it, particularly if you haven't been reading a lot of papers that are written this way, but one of the advantages of IMRD papers is that what goes where soon becomes obvious, both for readers and for writers. Readers of IMRD papers don't generally read them in order; instead they flip to the section that contains the information that interests them the most before deciding whether to read the rest. In the next few pages, we'll explain what goes where in an IMRD paper and why. Before we begin, we should note that in addition to the Introduction, Methods, Results, and Discussion sections, IMRD papers generally include abstracts. While abstracts are the first sections readers see in published papers, we recommend writing them last, so we'll talk about them last here.

To illustrate the various components of IMRD papers, we'll offer very short extracts from an undergraduate honors thesis called "In the Service of Which Master? Civil–Military Relations and Regime Collapse in the Arab Spring."

Introduction

The introduction to an IMRD paper needs to do the same work as the introductions we discussed in Chapter 3 and the introductions to research proposals we discussed earlier in this chapter: explain what your question is and why it matters, give your readers a sense of the scholarly conversation you'll be entering, and state your hypothesis if you have one. If the paper is a long one, such as the senior thesis we are discussing here, the introduction is likely to include a literature review. (We talked about how to write a good literature review in Chapter 3.)

Here is how the student sets up the *research question*:

The prevailing notion on Arab civil–military relations is that "on a political level . . . the armed forces' loyalties lie with the regime rather than with the general population, a democratic system or the nation as an abstraction," *but in certain instances [during the Arab Spring] the*

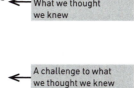

What we thought we knew

A challenge to what we thought we knew

military behaved in ways that totally defied the conventional wisdom. Thus, we are left with many questions without adequate scholarship to address them . . . most importantly, *how does the Arab Spring challenge the previous understanding of civil–military relations in the* ⟵ The research question *MENA [Middle East and North African] region?*

The portion of this student's literature review that describes past research (the scholarly conversation the student will be entering) describes past theories and findings on three topics: regime stability, civil–military relations, and protest repression.

Here's an example of how the student does that:

The most common theory that scholars offer [about regime stability] is ⟵ General theory *economic in nature.* The general line of thought is relatively simple: if a regime has money, it is able to appease its citizens via cooptation, quelling any desire for regime change. Investment in education, infrastructure, jobs, or other improvement that make the ⟵ Mechanism citizenry happy can be used. Regimes without money, however, are not able to co-opt the population and must instead rely on the citizenry's goodwill, which is less likely to work. Especially with respect to the Arab autocratic regimes, many scholars attribute regime stability to cooptation. *Sven Behrendt's work* on sovereign wealth funds— ⟵ Specific example of scholarship "investment funds that are owned or controlled by national governments"— exemplifies this line of thought.

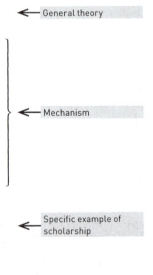

And Table 4.4 shows the student's hypotheses.

TABLE 4.4   **Student Hypotheses**

Hypothesis #1	Regimes that maintained a stronger civil–military relationship were most likely to survive the challenges presented by the Arab Spring.
Hypothesis #2	States that were willing to use violence at high levels against protesters were the regimes that collapsed, whereas those that did not use violence survived the challenge.
Hypothesis #3	High levels of violence resulted in military defection or fracturing, which in turn led to state collapse.
Hypothesis #4	Coup-proofing mechanisms played an instrumental role in dictating military response to the protests.

Methods

The difference between writing about others' methods and your own is that your own must be recounted in sufficient detail to allow somebody else to replicate them exactly. That means this section can be both the easiest and the least exciting to write because you simply describe everything you did in the order in which you did it. By "everything," here, we mean everything that matters. What matters may sometimes be an open question. For example, for some statistical analyses, the software and even the commands you used need to be reported, although you don't need to report whether you used Microsoft Word or Scrivener to write your paper. Similarly, for experiments, reporting the exact procedure and even the room setup may be important, but you generally don't need to report on what you wore to interview a subject. (Generally! However, if you conduct interviews at the homeless shelter, it may be sensible and therefore worth reporting that you dressed in informal and affordable clothing to fit in well with your respondents.)

Although we just said that the methods section is the most straightforward one to write, we should also note that the credibility of your work rides on your methods section, and some amount of justification must be folded into your description. For example, you might write, "In order to ensure consistency, all participants were interviewed using the same list of questions," which tells your reader both what you did and why you did it.

We mentioned above that each section of an IMRD paper plays a specific, prescribed role. In the methods section, you should only describe what you did. In fact, you may be noticing that we're using the word "describe" an awful lot, and there is a reason for that. You will have to resist saying anything about what you found, or how people responded to what you did: that must be saved for the results section.

Here's the overview of the methods used in the thesis about the Arab Spring, with the key methodological moves annotated:

To test these hypotheses, *this paper will use macro-level typologies to gain a better under-* ← Macro-level typologies
standing of the phenomena that occurred throughout the protests across the region. There is still considerable disagreement about what happened in the Arab Spring, so this data will hopefully provide a clear lens through which to understand, at least on the macro level, what occurred. *The paper will then grapple with the nuanced web of the civil–* ← Case studies
military relationship, use of state violence, defection and state collapse through case studies of Egypt, Tunisia, and Bahrain. . .

Table #1 will serve as the basis for the quantitative analysis section of this paper. Using typologies for Regime Type, Result of Protests, Civil–Military Relationship, Level of Violence, and Military Response, *it will cross-tabulate variables* in an attempt to see how each interacts and unpack the puzzle of

← Cross-tabulation of macro-level typologies with case study countries

how and why regimes collapsed or survived
and the role the military played in the collapse or the survival. The various typologies
for Table #1 were gathered from various
sources and require explanation.

The student then goes on to identify *data sources* and define
the *terms* used in the typologies. For example:

The "level of violence" typology uses the Uppsala Conflict
Data Program's PRIO database. "Low" indicates a death
toll of 1–24, "Intermediate" of 25–999, and "high" of 1000
or more. "None" is reserved for those countries that experienced Arab Spring protests but did not have any protest-
related deaths.

Results

While in the methods section you simply describe what you
did, *in the results section you describe what you found, without
offering your ideas about what your findings mean.* (There's that
word "describe" again.) For example, what answers did people
give in your interviews or surveys? What numbers did you
arrive at when you conducted your statistical analysis?

You might think of this section in terms of the saying from
an old television show, *Dragnet*: "Just the facts." By which we
mean, no interpreting.

The findings in the honors thesis about the Arab Spring are multifac-
eted, but here is the summary of a portion of the evidence:

First, it is clear that violence has a very
important effect on regime stability
throughout the Arab Spring, as only the

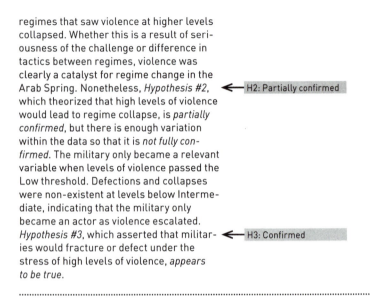

regimes that saw violence at higher levels collapsed. Whether this is a result of seriousness of the challenge or difference in tactics between regimes, violence was clearly a catalyst for regime change in the Arab Spring. Nonetheless, *Hypothesis #2*, ← H2: Partially confirmed which theorized that high levels of violence would lead to regime collapse, is *partially confirmed*, but there is enough variation within the data so that it is *not fully confirmed*. The military only became a relevant variable when levels of violence passed the Low threshold. Defections and collapses were non-existent at levels below Intermediate, indicating that the military only became an actor as violence escalated. *Hypothesis #3*, which asserted that militar- ← H3: Confirmed ies would fracture or defect under the stress of high levels of violence, *appears to be true*.

Discussion

In the discussion section, you finally get to interpret your results. What are the larger implications of what you've found? However, you need to state this carefully—using what we call "hedging" in Chapter 6—and you also need to consider alternative interpretations. As with the tendency to cherry pick, discussed in Chapter 3, we frequently encounter the desire to ignore ambiguity or to gloss over mismatches between theory and evidence. Such mismatches are, in fact, more the norm than the exception. Scholarly credibility arises out of modesty: admissions such as "evidence for my third hypothesis was not statistically significant," or "the fact that this survey's respondents were college students limits the generalizability of my findings," make you sound more open-minded, honest, and thorough than strong assertions that your data can only mean one thing.

In fact, all discussion sections should include two subsections—one on limitations, and one on directions for future research—whether or not these get singled out with subheadings (Table 4.5).

TABLE 4.5 **The Required Subsections of the Discussion Section**

Limitations	Any flaws, weaknesses, shortcomings, or barriers to your research. There is no such thing as a perfect research design, if only because none of us has infinite time, money, and knowledge. Thus, you must point out one or two of the most important things that could have made your research stronger.
Directions for future research	Suggestions about how to overcome the limitations you've described, and how to build on what you learned from the research you just conducted

Here are the suggestions for future research in the Arab Spring paper, with the possible approaches emphasized:

Future studies might consider examining the political relationship and makeup of [entities such as the military, internal security forces, and the police] attempting to gain a better understanding of their willingness to use force against protesters. Moreover, the elite security forces remain (intentionally) opaque and future research, if possible, *might look into illuminating their structure, recruitment methods, and doctrine.* As a general rule, Arab militaries do not publish the ethnic, religious, or social makeup of their troops. Analysts are forced to rely on sweeping statements such as "Bahrain's military is predominantly Sunni." Access would be difficult, but such a study would greatly enhance the political science community's ability to understand the military's actions in mass protests such occurred in the Arab Spring. Finally, almost all scholarship in response to the Arab Spring has examined the cases that saw Intermediate and High levels of violence, while those that experienced Low-No violence have garnered essentially no attention. *It is important that scholars examine what allowed the majority of countries not to escalate to higher levels of violence,* so that the international relations community might better understand how to avoid the death and destruction caused by situations such as Syria, Libya and, on a lesser scale, Egypt, Tunisia, Bahrain, and Yemen.

A common problem we see in the discussion section is that it pretty much repeats the results section. That's usually because the writer has not limited himself or herself to *just describing* the results in the prior section or because he or she forgets to say something about the broader implications, limitations, and directions for future research.

Abstract

We saved the abstract for last because these are incredibly difficult to write before you write the rest of your paper, but reasonably simple to write afterward. The abstract summarizes the content and explains the purpose of your paper in a single paragraph and contains only one or at most two sentences representing the content of each of the sections we described above. Your abstract should state the following:

1. Your research question and why it matters
2. The central components of your research methods
3. Your most important results
4. The meaning of your results

Abstracts generally do not include citations, except in rare cases where the purpose of the paper is to respond directly to a particular piece of past research, nor do they include direct quotations. They describe your own work in your own words, as concisely as possible. (See Chapter 6 for tips on how to make your writing more concise.)

Here is the abstract for the Arab Spring thesis, with these four elements annotated:

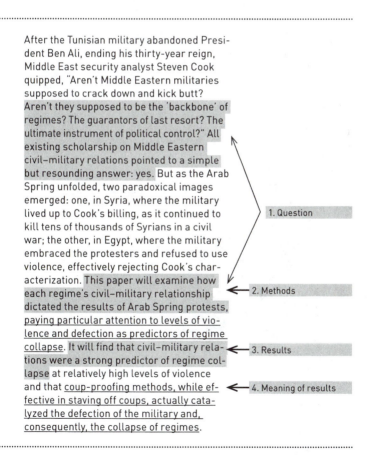

After the Tunisian military abandoned President Ben Ali, ending his thirty-year reign, Middle East security analyst Steven Cook quipped, "Aren't Middle Eastern militaries supposed to crack down and kick butt? Aren't they supposed to be the 'backbone' of regimes? The guarantors of last resort? The ultimate instrument of political control?" All existing scholarship on Middle Eastern civil–military relations pointed to a simple but resounding answer: yes. But as the Arab Spring unfolded, two paradoxical images emerged: one, in Syria, where the military lived up to Cook's billing, as it continued to kill tens of thousands of Syrians in a civil war; the other, in Egypt, where the military embraced the protesters and refused to use violence, effectively rejecting Cook's characterization. **1. Question** This paper will examine how each regime's civil–military relationship dictated the results of Arab Spring protests, paying particular attention to levels of violence and defection as predictors of regime collapse. **2. Methods** It will find that civil–military relations were a strong predictor of regime collapse **3. Results** at relatively high levels of violence and that coup-proofing methods, while effective in staving off coups, actually catalyzed the defection of the military and, consequently, the collapse of regimes. **4. Meaning of results**

The Problem of Too Much Information

In your everyday conversations, when you say, "Shut up! TMI!" you are usually responding to someone getting a little too graphic about issues of, say, personal hygiene, bodily functions, dysfunctional relationships, and the like. Scholarship also faces the problem of too much information. Although it is usually a bit less gross than the TMI in your everyday conversations, it is a potential problem.

We said earlier that you must describe all the relevant details about your methods and results. But what if you had a fifteen-page survey? What if your dataset had ninety variables? What if you have a separate graph for every county in California? The need for replicability requires total transparency, but you shouldn't kill your readers by boring them. Here is where appendices (that's the plural of appendix) can come in handy. Consider including an appendix in which you provide the complete information that your reader may want to consult, but that isn't immediately relevant to your central points. Alternatively—especially if you have collected your own data and created an original data set—you may also want to create a web repository for the supplementary information. Just make sure you include a prominent link to it in your text.

And speaking in general of any information that *might* be relevant but isn't directly germane to your argument, remember the humble footnote. It is an underused resource in undergraduate work. Footnotes don't have to be reserved only for citations; you can use them to flag related issues or literatures, to speculate, and, in general, to go on brief intellectual tangents. They can offer a way of increasing your credibility, by demonstrating that you know more about the topic than you have room to discuss.

Final Polishing: Titles and Revision

Once you've completed everything else, including your abstract, you'll want to give your paper a good title, and you'll want to revise it thoroughly to be sure it's the best it can be. See the final two sections of Chapter 3 for advice about crafting titles and revising your work.

CHECKLIST FOR DATA-DRIVEN RESEARCH PROPOSALS AND IMRD PAPERS

Do

✓ "Reverse engineer" published papers in your area to understand what kinds of methods and data to use.

✓ Propose research projects that are feasible for you, given your skills and resources.

✓ Use appropriate visuals to convey your results.

✓ Understand the structure of research proposals and IMRD papers, and stick to it.

Don't

✓ Use unnecessary, inappropriate or misleading visuals.

✓ Discuss the implications and limitations of your findings in the results section.

STRATEGIES FOR RESPONSE PAPERS, CASE STUDIES, ADVOCACY PAPERS, AND BLOG POSTS

As we said at the beginning of Chapter 3, you will need to use most if not all of the steps of the writing process we described there for a great many of your papers, including compare and contrast papers, literature reviews, research papers, and many of the apprenticeship genres your instructors create to help you learn specific things in their courses. However, we'd also like to provide you with strategies for writing certain types of papers that won't require you to use all of these steps, or that require a somewhat different approach.

Response Papers

Response papers are typically fairly short and are often used to help prime your thinking about something you've read. They are also often ungraded, but not always, and of course, you'll want to do a good job with them in any case. Sometimes, your instructor will provide a prompt that helps direct your attention to a particular aspect of the text you are being asked to respond to, but you may also simply be told to "write a one- or two-page response to the reading for today's class."

While a response should not be a summary, it should demonstrate that you understood the key ideas in what you've read. It should also be treated as a "real" paper, by which we mean it should have an introduction and conclusion, however brief these might be. One straightforward way to begin a response paper that makes it clear that you've understood what you've read is to write a single sentence that includes the author's name, the title of the text, and the main point of the argument. For example, "In 'Throwing Like a Girl,' Iris Marion Young argues that many of the supposedly physical differences in how women and men move are actually the result of internalized gender norms."

Most of the space in your response paper should be reserved for your response. If you haven't been given specific instructions about what kind of response to offer, you might either relate the main idea in this text to others you've already read in class, or write about how the main idea in the text does or doesn't apply to a political question that is relevant to your course. You might talk about what you find strange, interesting, or useful—keeping in mind that you are participating in a scholarly conversation. (In other words, unless you are specifically instructed otherwise, you should offer ideas, not opinions. If you're not sure what we mean by that, see our sidebar on "What's wrong with opinions?" in Chapter 2.)

Here are a couple of examples of the types of response papers we frequently see, but which *you should avoid writing*. First, **you should not focus on how challenging the text was** (unless that's the actual assignment):

I had a hard time with Iris Marion Young's text. I thought it was really dense.

You are not telling your readers anything really valuable about the text. (We are not discouraging you from telling your instructor if you struggle with course materials. We are just saying that the response papers are generally *not* for that purpose.)

And here is the kind of **opinion-based response that you should also avoid writing**:

> I liked Iris Marion Young's paper because I believe gender differences are not based in biology.

This response states a preference rather than an idea—the writer "likes" the paper because he or she agrees with it.

What should you do, then? Here's an example that offers **an idea related to the reading**:

> If Iris Marion Young is right, then we have all learned how to move in gendered ways, and we should also be able to learn to move differently; one question this raises is how much political effect moving differently would actually have.

This example suggests a strategy you might try if you're ever having difficulty coming up with a response of your own: rather than trying to figure out if you agree or disagree with what you've read, ask why and how the argument matters.

As with your introduction, your conclusion might be as short as a single sentence, but it must exist. It's fine if it's open-ended—that's the nature of this kind of assignment. For instance, you might suggest a research question the reading raises for you, or point out implications of your response that you don't have space to address.

CHECKLIST FOR RESPONSE PAPERS

Do

✓ Show that you've understood what you read.
✓ Include an introduction and a conclusion.

Don't

✓ Offer opinions instead of ideas.

Applying Theories to Cases

"Theories and the World" papers often ask you to consider how well a theory explains a particular case, or set of cases, in the real world. One goal of such a paper is to establish how well you understand the theory; another is to hone your analytic skills. Thus, your first task in a paper like this is to describe the theory you'll be working with as accurately as possible. Your second task is to think through how the features of your case (or cases) can or cannot be explained by the theory.

You may be asked to find your own case for a paper like this, or to consider the perspectives of critics of your theory, but for the most part, you won't have to do the kind of research that requires you to find outside sources. You may, however, be asked to consider how more than one theory applies to your case, or how a single theory applies to multiple cases. If so, we suggest that you look back at the section on organizing comparisons in the body of your paper in Chapter 3. If you are working with a single theory and a single case, it is usually most logical to explain all of the ways in which the theory does fit before explaining the ways in which it doesn't.

One mistake to avoid making in a paper like this is to assume that every aspect of the theory applies, and that you just need to figure out how. Your task is not to force your case to fit your theory, but to figure out what can be learned both from finding out what aspects the theory does account for and from what aspects it doesn't. Writing such a paper should help you understand your cases better. More interestingly, from the perspective of participating in the scholarly conversation, it should also give you insights into ways in which the theory might need to be revised. Such insights should help you answer the "So what?" question we discussed in Chapter 3.

CHECKLIST FOR APPLYING THEORIES TO CASES

Do

✓ Consider the ways in which mismatches between theories and cases suggest ways in which theories may need to be revised.

Don't

✓ Assume that every aspect of the theory applies to your case.

Advocacy Papers

Advocacy papers differ from other political science papers in terms of their purpose. In contrast to most other political science writing, advocacy papers are written to persuade readers to take specific action. In Chapter 1, we said that advocacy papers include both policy memos and op-ed articles, and that one key difference between them is the target audience: for policy memos, your audience probably already has quite a bit of expertise in the area you are writing about, while for op-eds, your audience is more likely a general reader with little to no expertise. A policy memo might advise a member of the State Department about military or economic intervention in another country. An op-ed might try to convince the general public to favor one type of intervention over the other and email their congressional representatives about the matter.

Advocacy papers share many features with more conventionally "academic" papers. The most important of these is that you'll need to make your main point early. Real-world policy memos frequently have a very short section called the "executive summary." The idea is that real decision makers—the executives—won't have time to read the whole memo. They simply need to know what needs to be done, and why. Whether your assignment explicitly asks you to write a section called "executive summary" or not, we recommend beginning as if you are writing one. This can be very short:

> The violent persecution of the yoighur minority in Bajikistan has increased by 200% over the last year and risks becoming a genocide. Given both the strategic significance of Bajikistan for the United States *and* the importance of fostering better human rights in the region, we advocate economic sanctions against Bajik political elites and an increased military presence in the region.

Successful op-ed pieces often follow this same convention. Even though op-ed articles are not the same as news articles— the "op" stands for "opinion"—they follow the journalistic convention of getting to the point right away.

Whether you are writing a policy memo or an op-ed article, the rest of your advocacy paper then makes the case for the policy you recommend in the "executive summary." Although the specific structures of such papers can vary, depending on how much space you have—and on your instructor's specific requirements, of course—the following are generally important elements:

- *Why does this issue matter?* Unlike in the conventional academic essays we discussed in Chapter 3, here the answer to the "So what?" question is purely about the real world. The Secretary of State or readers of the *Fresno Bee* don't care whether your contribution engages a debate in political science.
- *What reasons and evidence support your recommendation?* Because the audiences for these types of papers vary, the way you use, discuss, and even cite evidence usually differs

from more academic papers—but you still need to back up what you claim. Academic citation conventions are often acceptable in policy memos, but not in op-ed articles. (Have you seen many newspaper articles with footnotes or bibliographies? We, neither.) In op-ed articles, you can simply mention authors and sources in the "body" text: "Political scientist Robert Axelrod has argued . . ."

- *How should your data be presented?* Policy memos often include data in the form of tables or figures, just like conventional academic work, whereas op-ed articles have these less frequently. And, given the audience, neither a policy memo nor an op-ed would be likely to include a mathematical model or a regression analysis table (see Chapter 4 for info on tables and figures).

- *What would someone who disagrees with you say, and how would you respond?* Considering counterarguments is particularly important in advocacy papers. In policy memos, it is important because decision makers need to be confident that all reasonable alternatives have been considered. In the world of op-ed writing, articles get published *because* they take a position on a controversial issue: you write because you know someone disagrees with you. You will lose credibility if you pretend this is not the case.

The labels "advocacy" and "opinion" might seem to invite you to do exactly what we counseled against in the section on "What's wrong with opinions?" in Chapter 2. But they don't. Advocacy papers want advocacy and claims based on reasons and evidence, not on your personal likes or dislikes. For this reason, it is also helpful to keep your tone dispassionate, however passionate you might feel about the issue. Advocacy papers are not political pamphlets. Op-ed articles allow for a bit more room for passion than policy memos, but it never helps your credibility if you come across as a rabid firebrand who is too angry to think straight.

CHECKLIST FOR ADVOCACY PAPERS

Do

✓ State your main point right away.
✓ Take counterarguments into account.

Don't

✓ Offer opinions instead of ideas (even in an op-ed).

Blog Posts

Blogs can play many different roles in courses. They can serve as a platform for response papers or other kinds of brainstorming and discussion among the students in a course. Here, we'll focus on assignments that ask you to participate in the real blogosphere or that at least treat blogging as if you were doing that. In other words, we'll discuss the blogosphere as a kind of genre.

Of course, there is no single blogosphere; there are fashion blogs, sports blogs, travel blogs, medieval studies blogs, and political science blogs, all with slightly different conventions. The first thing you'll want to know with a blogging assignment, therefore, is what kind of blog is expected of you, and whether your instructor might be able to point you to examples. Maybe the instructor hopes for something like The Monkey Cage (www.washingtonpost.com/blogs/monkey-cage/), which is now syndicated by *The Washington Post* and to which professional political scientists contribute actively. Or maybe your instructor wants the blog to have more of the "political" and less of the "science," as you might see on such high-profile blogs as Daily Kos (www.dailykos.com) or Politico (www.politico.com). Whatever the specific expectations, the following points are generally true of academic blogs in political science:

- *Blogs are partly a visual medium.* You therefore need to think about incorporating visual elements, whether photos and other pictures, or data and figures. (Be careful about intellectual property considerations, though, as we'll note later.) You will also need to learn how to embed images in a way that enhances the visual appeal of your post, instead of distracting from it. For example, you don't just want to center your images like Fig. 5.1,

FIGURE 5.1 A Poorly Placed Image in a Blog Post.

with a lot of white space around it. Instead, you will want to vary their alignments between left and right (and maybe centered), and let the text "wrap" around them, as we do with Figure 5.2:

FIGURE 5.2 A Better Location for an Image in a Blog Post.

- Even academic blogs *usually cite their sources with embedded hyperlinks instead of footnotes, in-text citations, or bibliographies.* Blog posts often enter their particular conversations by invoking other bloggers or news stories near the beginning with an embedded hyperlink, and they also cite their other sources with hyperlinks. You will need to embed those links into the words you use instead of pasting URLs directly into the text. Because this is a book that does not include hypertext, we had to include The Monkey Cage's URL in parentheses; if this were a digital medium, we would have embedded the URL into the words "The Monkey Cage." When you cite a source that is not directly accessible digitally—a book, for example—it is common to provide a link to the book's entry on a library website or an e-commerce site.

- If the blog you contribute to is public—and therefore open to your parents and maybe your future employers to read!—remember that the readers don't know what has happened in your course unless you tell them. So don't say things like "As Professor Schnitzel pointed out last week . . ." or "Young's argument seems weird." Instead, inform your audience: "Last week in my political science course, Professor Schnitzel introduced the concept of the collective action problem" or "In her essay 'Throwing Like a Girl,' Iris Marion Young argues . . ." and *then* go on to say what you meant to say.

- Many instructors treat blogging as a "low-stakes" assignment: as with response papers, the goal may be just to get you thinking. Posts may therefore be graded as pass/fail or with less attention to the nitty-gritty details of sentence structure and prose. This can be very helpful, both for you and for the instructor; we certainly think low-stakes assignments are important and have used them regularly. But especially if the blog to which you contribute is public, remember you'll still want to put some effort into the standard practices of good writing: think about what to say before you start, revise if needed, and proofread what you've written.

- Even in the blogosphere, and certainly in the academic blogosphere, *reasons and evidence* are more valued than the mere expression of likes and dislikes. Yes, we know the blogosphere is full of ranting, personal attacks, and trolling (deliberate attempts to disrupt conversations and demean individuals or ideas). If your instructor uses a blog in your course, it is very unlikely that he or she is interested in having you practice your trolling skills. Blogs certainly allow for more personal perspective and opinion than your political science papers or even advocacy papers, but personal opinions are rarely the point of the blog assignments. This may seem odd in political science—isn't much of our political discussion all about personal opinions and beliefs?—but we nevertheless recommend the same policy we advocate for all kinds of writing in political science: what matters most is not *what* you believe (you are welcome to hold any view) but *what reasons* make the belief reasonable, interesting, or worthwhile.

A Caution about Blogs and Intellectual Property

We live in a culture of remixing, sampling, and viral memes: borrowing others' ideas, images, and snippets of text or music, especially on the Internet, is seen as fair game. Many of us think such a culture of borrowing can be interesting, even great, but legally, it is a murky terrain. Legally, many acts of digital borrowing are *not* cases of fair use. Fair use is a legal principle that regulates whether and how someone may use someone else's intellectual property.

This adds a complication to course blogs. We said earlier you should use visual elements in your blog posts. What images can you use? Strictly speaking, you may only use material for which you are the intellectual property owner (by which we mean creator) *or* that you have permission to use. If an image is copyrighted or is on a page that is copyrighted—that is, you'll see somewhere that familiar © symbol—the only way you may use it on a public blog is by writing to the copyright owner and requesting permission. How likely are you to do that for your course blog post, due tomorrow morning? If you are like most of the students (and people) we know, not likely at all.

So what can you do to make your posts lively and not get you, your instructor, or even your college in trouble? Here are some suggestions:

- Use your political science blog as an opportunity to practice your artistic side: use your own photographs or illustrations. (Remember, though, that if you take a photograph of a copyrighted item, you have a copyright to your photo, but the item's owner retains his or her copyright to that item.)
- Use images that are in the public domain. Most of the material in the Library of Congress (www.loc.gov), for example, is in the public domain and free to use. Google's image search allows you to specify a search parameter that only returns items that are free to use. (Don't trust it blindly, though: double-check that the image you are interested in really is not under some license.)
- Use items licensed under the Creative Commons (creativecommons.org). Creative Commons is a licensing regime that is an alternative to the older idea of copyright. It is explicitly meant for our "culture of borrowing" and is therefore more flexible than copyright. When you see an item accompanied by something like Figure 5.3, you'll know it is licensed via Creative Commons and possibly useable. A Creative Commons license does *not* mean you can use the item any way you want; the particular license code will indicate how you may use it, and the Creative Commons website explains the license codes. In every case, at a minimum, you must credit the item's license owner.

FIGURE 5.3 A Creative Commons License Logo.

The folks at Creative Commons have the right idea: even when something is free to use, do be sure to give credit to where you got it. That's a good legal, ethical, and intellectual practice.

CHECKLIST FOR BLOG POSTS

Do

✓ Include visual elements.
✓ Use links to cite your sources.

Don't

✓ Write anything you wouldn't want your parents or a future employer to read.
✓ Use copyrighted images without permission.

STYLE IS MEANING

If you've been taking classes in different types of disciplines recently, you've probably already noticed differences in writing style between the humanities, social sciences, and natural sciences. Yet, while stylistic contrasts between a chemistry lab report and a comparative literature paper ought to be obvious to almost anyone, smaller stylistic differences also exist within these broader disciplinary categories. In this chapter, we'll outline common stylistic conventions in political science. We'll also offer advice about crafting clear prose and describe some common missteps you'll want to avoid.

Signposting

Although students sometimes think they need to build up to their main point by presenting all the information necessary to support or explain it first, political scientists do not appreciate being kept in suspense: they want to know the contribution you will make to the scholarly conversation right up front. Stylistically, most political science writing engages in consistent signposting: authors achieve clarity by signaling their intentions early and frequently. A "signpost" is exactly what it sounds like: it lets your readers know what direction your paper will be going in and announces any detours you might take along the way

(Fig. 6.1). This means you need to make your thesis or hypothesis clear very early on—ideally in your first paragraph, if your paper is a short one, and certainly by the end of your introduction. It also means you should never leave a reader wondering if you are ever going to get around to talking about a key idea or a major objection to your argument. For example, the last two sentences of the first paragraph of this chapter are a signpost that tells you what you'll learn here if you read on (common stylistic conventions, how to craft clear prose, and missteps to avoid).

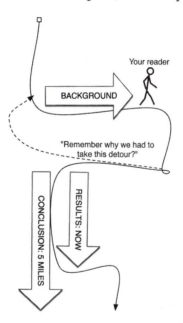

FIGURE 6.1   Signposts Keep the Reader Aware of Where the Paper is Going and Why

A **signpost** lets your readers know what direction your paper will go in and announces any detours you take along the way.

If you flip back to the first page of each chapter in this book, you'll see that it includes a signpost that tells you what you can expect to find in that chapter. At various other points, we also let you know that you can learn more about certain topics by turning to other chapters; for example, at the end of the section called "Common genres of political science writing" in Chapter 1, we mention that we'll be talking more about what we call "apprenticeship genres" in Chapter 2.

Signposts can point back toward earlier chapters as well. These often begin with phrases such as, "As we mentioned in Chapter 1 . . ." and their purpose is to remind readers about concepts or ideas that have already appeared in the paper without having to go into detail about them all over again. Backward-pointing signposts can also reassure readers that you are aware that you've already talked about some topic when you bring it up again, so they won't worry that you're going to repeat yourself or that the rest of your paper will be disorganized. It's a way of letting readers know that you're being careful about the route you've planned out for them, and that you are keeping track of every step along the way to be sure they won't get lost.

We've noted that writing is a recursive process and that you should expect much of what you plan to say to change while you write. Signposting can reveal a lack of revision if you're not careful. (You may say that you'll address an issue "in the next section" in your introduction, but find that it doesn't appear until three sections later when you go back and revise the paper.) Make sure your signposts line up with the actual content of your work before you hand your paper in.

Transitions

Like signposts, transitions guide your reader by signaling connections between what you've already said and what you'll talk about next. They typically occur at the beginnings of paragraphs and the beginnings of sentences. (Headings, too, serve as large and obvious transitions that help readers keep track of your argument. They can be particularly useful in longer papers.)

A good transition describes the direction of your thoughts; it lets the reader know whether you are developing an idea, qualifying it, or moving on to something new. "Furthermore," for example, implies that you are building on or strengthening a claim you've just made, while "however" indicates that you are raising a caveat or objection. "Alternatively" tells the reader you are about to address a whole new possibility.

The following paragraph from a senior thesis offers clear transitions between each sentence:

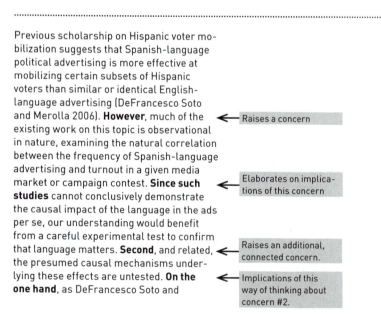

Previous scholarship on Hispanic voter mobilization suggests that Spanish-language political advertising is more effective at mobilizing certain subsets of Hispanic voters than similar or identical English-language advertising (DeFrancesco Soto and Merolla 2006). **However**, much of the ← *Raises a concern* existing work on this topic is observational in nature, examining the natural correlation between the frequency of Spanish-language advertising and turnout in a given media market or campaign contest. **Since such** ← *Elaborates on implications of this concern* **studies** cannot conclusively demonstrate the causal impact of the language in the ads per se, our understanding would benefit from a careful experimental test to confirm that language matters. **Second**, and related, ← *Raises an additional, connected concern.* the presumed causal mechanisms underlying these effects are untested. **On the** ← *Implications of this way of thinking about concern #2.* **one hand**, as DeFrancesco Soto and

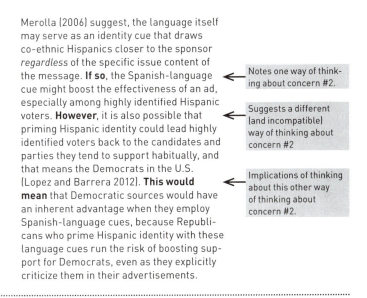

Merolla (2006) suggest, the language itself may serve as an identity cue that draws co-ethnic Hispanics closer to the sponsor *regardless* of the specific issue content of the message. **If so**, the Spanish-language cue might boost the effectiveness of an ad, especially among highly identified Hispanic voters. **However**, it is also possible that priming Hispanic identity could lead highly identified voters back to the candidates and parties they tend to support habitually, and that means the Democrats in the U.S. (Lopez and Barrera 2012). **This would mean** that Democratic sources would have an inherent advantage when they employ Spanish-language cues, because Republicans who prime Hispanic identity with these language cues run the risk of boosting support for Democrats, even as they explicitly criticize them in their advertisements.

> Notes one way of thinking about concern #2.

> Suggests a different (and incompatible) way of thinking about concern #2

> Implications of thinking about this other way of thinking about concern #2.

You may have noticed that this writer used the word "however" twice in the paragraph above. That's fine, but if that happened two or three times on every page, we might begin to find the repetition monotonous. In Table 6.1, we suggest variations for some transitions commonly used in political science.

TABLE 6.1 **Commonly Used Transitions**

However, yet, despite this, nevertheless	Suggest reservations, concerns, or caveats
Clearly, indeed, certainly, in fact	Show emphasis
For example, for instance, in particular, specifically	Introduce an example
On the other hand, in contrast, alternatively	Introduce an alternative
This suggests, this implies, therefore, thus	Introduce implications
Similarly, in the same way, likewise	Suggest similarity

Hedging

Authors in political science achieve credibility through judicious stance-taking. While it is important to articulate clear claims, it is also crucial not to overstate the certainty of your claims or your findings. Unlike mathematics, political science is not a field in which one can "prove" things (with the single exception of "proving" the internal consistency of a mathematical model of politics), and political science tends to be marked by more "hedged" claims than writing in other disciplines.[1] To "hedge" a claim means to limit or weaken it, and perhaps ironically, in political science the strength of one's argument frequently increases with the weakness—or at least modesty—of one's assertions. Claims that are asserted too strongly sound less objective; if the writer is unwilling to countenance doubt, the argument seems more like a product of his or her beliefs or political commitments than a testable theory. The differences between common stance-taking words and phrases are often subtle and may be hard to detect at first (and even more so if your first language doesn't happen to be English). Next we will work through some examples.

While claim-making generally occurs throughout a paper, there are **three key claim-making moments** we'll want to focus on here:

1. Presenting a thesis or hypothesis in your introduction
2. Differentiating your approach from those others have taken
3. Explaining your conclusions

Introductions and conclusions are the places where inexperienced writers are most likely to assert that they "will prove" or "have proven" something. "Proof" is far too strong a term to apply to the kind of patterns we can discover in political and social phenomena; it implies the impossibility of deviation. Such strict regularity exists only in relationships between mathematical and logical concepts, not in relationships between human beings.

What, then, might you be able to accomplish? You might be able to "show" or "demonstrate" that a pattern or relationship has existed in more than one setting (multiple time periods or several countries, for example). It may not sound like it to you, but both "show" and "demonstrate" indicate a pretty strong stance. If you say you've demonstrated something, you're asserting your conclusion fairly confidently. A safer verb to use, and one that might make you less of a target for the irritation of scholars who disagree with you, is "suggest." If you say your findings "suggest" you've discovered a new phenomenon that others should pay attention to, you're taking a modest stance; you're pointing out a possibility rather than a certainty (Table 6.2).

TABLE 6.2 **Verbs and Hedging**

Least Hedged ⟵			⟶	Most Hedged
~~prove~~	demonstrate show	appear to demonstrate may show	suggest	may suggest

Of course, all of these options can be hedged further: "the data appears to show" and "the data seems to demonstrate" are obviously less assertive claims than "the data shows" and "the data demonstrates." Similarly, "this may suggest" is a more cautious variation of "this suggests."

They can also be bolstered: "this strongly suggests" is a bit more aggressive than "this suggests," and "the data clearly shows" is quite a strong pronouncement.

Differentiating your approach or position is a tricky business. You don't want to insult anyone, after all—at least, not usually. Hedging can help you avoid that. Suppose you want to take on an argument that you believe is completely incorrect. Generally speaking, stating "X, Y, and Z are completely incorrect" will not win you any points, either with X, Y, and Z, or with your readers. Instead, you might try something like this: "X, Y, and Z have suggested that oranges are the most politically interesting fruit. Yet, while they have compared oranges, apples, and bananas, kiwis remain to be tested. A direct comparison of oranges and kiwis may offer a new perspective on this question." An experienced reader will understand that what you mean by this is, actually, "X, Y, and Z are completely incorrect" but will admire the skill and good manners you show in stating it less directly (Table 6.3).

TABLE 6.3   Different Ways of Signaling Disagreement

Insultingly Direct Way to Disagree	Diplomatic (hedged) Way to Disagree
"X is completely incorrect regarding oranges."	"X has suggested that oranges are the most politically interesting fruit. Yet, other fruit remains to be tested."

Let's look again at part of the paragraph from the senior thesis we studied earlier to see how it makes the same kind of move:

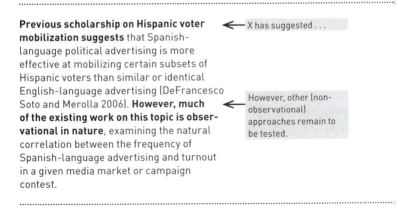

Previous scholarship on Hispanic voter mobilization suggests that Spanish-language political advertising is more effective at mobilizing certain subsets of Hispanic voters than similar or identical English-language advertising (DeFrancesco Soto and Merolla 2006). **However, much of the existing work on this topic is observational in nature**, examining the natural correlation between the frequency of Spanish-language advertising and turnout in a given media market or campaign contest.

← X has suggested . . .

← However, other (non-observational) approaches remain to be tested.

As you can see, a properly hedged response to past research depends on careful choice of both verbs ("suggests" rather than "claims to have shown") and transitions ("however" or "yet" rather than "on the contrary"). If you find yourself struggling to be diplomatic, imagine that the writer whose work you're responding to will read your paper someday. And if that still doesn't do it, imagine that he or she will read your paper right before having to decide whether or not to offer you a job.

Crafting Clear Prose

In this section, we'll spend some time on how to write clear prose. We don't have time to cover everything a person might ever want to know about writing clearly, of course, so the two components we'll focus on here are helping a reader focus on your point through context-setting and emphasizing new information, and avoiding vagueness by choosing specific verbs. We know many student writers have questions about the difference between active and passive voice, so we'll explain that as well. Building your sentences well and choosing your verbs with precision can contribute greatly to the clarity of your writing.

Context and Emphasis

The easiest way to confuse a reader is to fail to set the context for your arguments. We haven't used the language of "context-setting" in our discussion of writing introductions, but that is what an introduction does: it provides the background information a reader needs to understand what a paper is about and why it matters. In contrast, emphasis lets a reader know what is new and important about what you're studying and writing about.

According to George Gopen and Judith Swan,[2] readers can most easily absorb what you have to say when you set the context at the beginning (of a paper, a paragraph, or a sentence) and emphasize new information by placing it at the end. This is why you answer the "So what?" question in your conclusion: doing so allows you to emphasize your contribution.

To see how context-setting and emphasis work, let's revisit a sentence from the introduction to the senior thesis we looked at in Chapter 4:

The prevailing notion on Arab civil–military relations is that "on a political level . . . the armed forces' loyalties lie with the regime rather than with the general population, a democratic system or the nation as an abstraction," **but in certain instances [during the Arab Spring] the military behaved in ways that totally defied the conventional wisdom.**

⭠ The old, familiar idea about Arab civil–military relations = background you need to know

⭠ New information: emphasized because it needs to be explained

---

If the student had written this the other way around, this sentence not only would have been confusing but also would have lost a lot of energy. Consider this variation:

The military behaved in ways that totally defied the conventional wisdom during the Arab Spring, which was unexpected given the prevailing notion that "on a political level . . . the armed forces' loyalties lie with the regime rather than with the general population, a democratic system or the nation as an abstraction.

⭠ Here the reader may be thinking, "Wait! What conventional wisdom?"

⭠ Now the reader is focused on this idea, which is old news, instead of on the interesting question the writer has raised.

This idea applies as you move from one sentence to the next throughout your paper: old information a reader will need to understand what follows belongs at the beginning, and new information or ideas you want to emphasize belong at the end.

Choosing precise verbs

Like sentence structures, verbs play a big role in a reader's ability to understand what a writer wants to convey. For example, consider the difference between the verb "wrote" and the alternatives listed in Table 6.4.

TABLE 6.4 **Alternatives to "Wrote"**

wrote	argued
	analyzed
	criticized
	described
	debunked
	promoted
	ridiculed

"Wrote" really just means that a writer produced the content of a text. It doesn't tell us anything about that writer's purpose for doing so.

A paper that begins something like, "In *Leviathan*, Thomas Hobbes wrote about the social contract," is off to an unnecessarily slow start. This writer may as well have written, "There is stuff about the social contract in Hobbes's book." What kind of stuff did Hobbes write? We must wait for a whole new sentence to find out. Much clarity and efficiency can be achieved by selecting a more specific verb. For example, "In *Leviathan*, Thomas Hobbes argued on behalf of the social contract."

In addition to helping a reader understand the agenda behind another writer's work, verbs can signal your stance toward another writer's ideas. "Asserts," "states," "argues," "implies," "claims," "suggests," and "notes" are all verbs that can be used to tell a reader what somebody else said. (As in, "Jones asserts/states/argues/implies/claims/notes that increasing globalization has had a detrimental effect on several transportation industries.") However, they carry different evaluations of the ideas being conveyed (Table 6.5).

TABLE 6.5 **Verbs Signaling Different Stances**

Positive stance (takes ideas as fact)	notes
Neutral stance	states argues suggests
Slightly negative stance (implies skepticism)	asserts claims implies

You may be surprised to hear that "notes" signals agreement and "implies" signals a raised eyebrow, but it should be clear from the earlier section on hedging that you are unlikely to encounter a stance-taking verb that is more than slightly negative or minimally positive.

Active versus Passive Voice

You may have learned in your English composition courses that passive verbs (or "passive voice") should be avoided, and indeed, in many cases, active verbs create more efficient sentences. Yet, there are times when passive verbs are the appropriate choice. To help you sort this out, we should make sure you know what does and does not constitute a passive sentence construction. Many people think any sentence with a "to be" verb in it ("is," "was," "are," or "were") is passive, but this is not correct. Passive sentences make objects their subjects; things are done to someone or something, instead of someone or something doing something. Hence the word "passive." Here are a few examples:

Ten respondents were interviewed. (Passive)
We interviewed ten respondents. (Active)

Note that doer of the action (we) is missing from the first version of the sentence above. In contrast, consider the following sentence:

Ten respondents were hungry.

This sentence uses a "to be" verb (were), but the sentence is not passive; the respondents are not the objects of a missing agent's action. One way to tell a passive sentence from one that simply happens to include a "to be" verb is to think about whether it would make sense to add "by" to the sentence. ("Ten respondents were interviewed **by the researchers**" makes sense; "Ten respondents were hungry **by the researchers**" does not.)

Expectations regarding the use of passive voice are discipline-specific: writers and readers in some natural science fields regard it as more appropriate to leave the researchers out of a discussion as much as possible to keep the focus on what happened and what was found rather than who made it happen and who found it, which practitioners in these fields regard as sounding more objective. In these cases, you will encounter sentences that say "An experiment was conducted . . ." rather than "We conducted an experiment . . ." or "The planets were observed to align . . ." rather than "We observed that the planets aligned . . ."

Political scientists tend to prefer active constructions. However, at moments when what has been done is notably more important than who did it, passive voice may be appropriate, particularly if you are writing the kind of IMRD-style paper we discussed in Chapter 4. For example, "The trend was found to be significant at the .001 level" highlights the issue of significance. The reader already knows that the researchers did the significance testing, so including that information isn't important.

Avoiding Common Missteps

In the final section of this chapter, we'd like to help you avoid making certain moves that we'll call "common missteps": asserting grandiose claims (trying to make your argument sound more important than it is), overwriting (trying to sound more erudite than you are), and engaging in unintentional sexism (writing as if everyone on the planet is male). We'll address each of these separately.

Grandiose Claims

Students often try to signal the importance of their arguments by making overly broad claims in their introductory paragraphs ("Since times immemorial, people have argued over the true meaning of justice," or "In our globalized world, nothing is more important than public health"). First, you should be aware that both of those formulations (and many others, such as "In today's society . . .") are clichés. Your instructor has seen hundreds of papers that begin with sentences like those. Second, neither of those sentences tells the reader anything specific enough to be interesting.

If you think in terms of the scholarly conversation we've referred to throughout this book, you'll see that the first of those sentences is also irrelevant. We don't care what people have argued about "since times immemorial." We care about what they are arguing about right now. And there's a way in which the second version is rude. It doesn't say, "I have something interesting to contribute." It says, "My topic is the most important topic in existence." That type of claim is pretty hard to defend. You'd have to muster astonishing, indisputable evidence to support it.

Rather than beginning a paper with a grandiose claim, which is sort of like marching into a party in a purple suit and speaking through a bullhorn, try for something more manageable and more accurate—not "Since times immemorial, people have argued over the true meaning of justice," but perhaps,

"Recent Supreme Court cases have brought the relationship between justice and punishment into question." And instead of "In our globalized world, nothing is more important than public health," you might try, "Given the current ease and frequency of international travel, devising rapid and effective ways of dealing with deadly diseases such as Ebola is crucial to avoid both cultural panic and unnecessary deaths." In other words, aim for claims that give your reader an accurate sense of the scope and content of your argument (Table 6.6).

TABLE 6.6 Grandiose Versus Manageable Claims

Grandiose and Vague Claim	Manageable and Specific Claim
"Since times immemorial, people have argued over the true meaning of justice."	"Recent Supreme Court cases have brought the relationship between justice and punishment into question."
"In our globalized world, nothing is more important than public health."	"Given the current ease and frequency of international travel, devising rapid and effective ways of dealing with deadly diseases such as Ebola is crucial to avoid cultural panic and unnecessary deaths."

Responding to a specific writer or set of writers might also help you generate a more manageable claim. Instead of those vague "people" who have argued "since times immemorial," you might respond directly to a recent theorist such as John Rawls (Table 6.7).

TABLE 6.7 Vague Versus Precise Claims

"Since times immemorial, people have argued over the true meaning of justice."	"John Rawls's argument that the principles of justice must be compatible with those that people would choose if they did not yet know what their own real-world class position or social status would be poses the difficulty of deciding what values a person without any social characteristics might hold."

As you can see, in each of our modified examples, we suggest that you begin with a claim that applies specifically to the paper that will follow rather than one that might just as easily be pasted into any number of other papers focusing on questions only vaguely related to your own.

Overwriting

It is common for students to try to signal erudition by using words and sentence structures they perceive to be "scholarly" but that in fact result in convoluted, confusing, and unintentionally humorous prose (Fig. 6.2). We understand that this can be part of the learning process; if a writer has been told to "write like a political scientist," or in other words, like a professor, it's not surprising if he or she makes an effort to sound "professor-like." However, we recommend that you try to sound like the most easily understood professor you know. Stuffiness is not a virtue, in political science or anywhere else.

FIGURE 6.2   Calvin thinks he has figured out the art of academic writing. Don't be Calvin! Calvin and Hobbes © 1993 Watterson. Reprinted with permission of Universal Uclick. All rights reserved.

Two principles to help you avoid overwriting:

1. **Choose words of few rather than many syllables** ("use," not "utilize"), unless the longer word is

continued

> truly more precise than the shorter one ("experi-ment" vs. "test").
> 2. **The more complicated the idea you are trying to explain, the simpler your sentence structures should be.** Don't challenge your reader (and yourself) on both fronts at once.

We should say more about that second point on sentence structures. Though you are likely to read many very long and convoluted sentences during your college career, you should also be aware that conciseness is a virtue in academic writing (and writing in general). Sometimes students use more words than they need when trying to "fill out" papers they think are too short. (Trust us—your instructors can tell when you do that.) But we know that sometimes they also simply develop bad habits.

Here are four quick tips for producing more concise prose:

1. **Eliminate unnecessary prepositional phrases**
   - "In the book that was written by Fukuyama" → "In Fukuyama's book"
   - "The reason for the failure of this policy" → "This policy failed because"

2. **Eliminate redundancy**
   - "The people were motivated to be mobilized to vote." → "The people voted."
   - "The local citizens living in the area turned out to vote."→ "The local citizens turned out."

3. **Eliminate "There is/are" and "it is" constructions**
   - "There are two hundred refugees who need political asylum." → "Two hundred refugees need political asylum."

- "It is the capital city that stands to gain the most from this aid package." → "The capital city stands to gain the most from this aid package."

4. **Omit "filler" phrases that add no additional meaning**
   - "All things considered, this new policy has improved the situation." → "This new policy has improved the situation."
   - "For all intents and purposes, this new policy has improved the situation." → "This new policy has improved the situation."

Perhaps the simplest rule we can offer you for writing concise sentences is to make every word matter. If eliminating a word or phrase won't change the meaning of your sentence, take it out.

Unintentional sexism

You are probably aware that it is not acceptable to write about the hypothetical individuals in your papers ("the average American voter," for example) as if they are all men. However, you have probably also noticed that repeating "he or she" and "his or her" all the time begins to sound pretty awkward. You may be tempted to substitute "they" or "them," even when referring to individuals rather than groups. While we admit that this has become common in American speech—we ourselves occasionally speak that way in casual conversation—and writing as well, many instructors believe substituting "they" for "he" is not appropriate in academic writing, so we discourage that approach. It is fine, however, to change your overall example so that you refer to "average American voters" and use the plural form consistently. We'll discuss a variety of alternatives here.

Here's a problematic example that we'll want to repair:

"When the average American voter goes to the polls on voting day, he is typically familiar only with the most highly publicized electoral races, and when he is confronted with choices between candidates for more obscure positions, such as nonpartisan judicial candidates, he must decide whether to cast his vote in ignorance or to abstain."

What strategies are available to you for rewriting this sentence so that it doesn't suggest that all voters are male? Here's a list of possibilities:

Strategies for Replacing "He"

1. Use "he or she."
2. Make the subject of the example plural so you can use "they."
3. If you've used a noun in your example (voter, politician, etc.), continue to use the noun rather than shifting to pronouns.
4. Restructure the sentence to eliminate the need for either nouns or pronouns.

Use "he or she" and "his or her"
We can certainly change "he" to "he or she" throughout this example, but the result isn't particularly elegant:

"When the average American voter goes to the polls on voting day, **he or she** is typically familiar only with the

> most highly publicized electoral races, and when **he or she** is confronted with choices between candidates for more obscure positions, such as nonpartisan judicial candidates, **he or she** must decide whether to cast **his or her** vote in ignorance or to abstain."

Ugh. There is nothing grammatically incorrect about that sentence, but the repetition of the three-word phrase "he and she" or "his or her" becomes annoying. Let's try option number two.

Make the Subject Plural

Note that you'll also need to change your verbs to the plural form when using this approach, and perhaps a few other words as well. Sometimes this also entails deleting words, like the word "the" before "American voters." We've highlighted all of the changed word forms in the example below.

> "When average American **voters** go to the polls on voting day, **they are** typically familiar only with the most highly publicized electoral races, and when **they are** confronted with choices between candidates for more obscure positions, such as nonpartisan judicial candidates, **they** must decide whether to cast **their votes** in ignorance or to abstain."

This sounds fine; "they" is much less cumbersome than "he or she" when it must be repeated several times.

Stick to Nouns Instead of Pronouns

Sticking with nouns is often viable as well, although nouns must often be accompanied by articles ("the" or "an"), and this can end up sounding just as clunky as using "he or she," or even worse, as in the following example:

> "When the average American voter goes to the polls on voting day, **the voter** is typically familiar only with the most highly publicized electoral races, and when **the voter** is confronted with choices between candidates for more obscure positions, such as nonpartisan judicial candidates, **the voter** must decide whether to cast **the voter's** vote in ignorance or to abstain."

"The voter must decide whether to cast the voter's vote"? Ack!

Some Nouns Removed
You may have been itching to delete a few instances of "voter" while reading that last version. If so, good for you. Let's see how that would look. We'll mark our deletions with brackets if we don't replace them with anything else.

> "When the average American voter goes to the polls on voting day, the voter is typically familiar only with the most highly publicized electoral races, and when [] confronted with choices between candidates for more obscure positions, such as nonpartisan judicial candidates, [] must decide whether to cast **a** vote in ignorance or to abstain."

Combined Strategies
Combinations of these approaches are also possible, and often desirable, to help avoid monotony. Here's an example that uses

three out of the four strategies suggested above (using "he or she," using nouns instead of pronouns, and eliminating the need for either a noun or a pronoun).

"When the average American voter goes to the polls on voting day, **he or she** is typically familiar only with the most highly publicized electoral races, and when [] confronted with choices between candidates for more obscure positions, such as nonpartisan judicial candidates, **the voter** must decide whether to cast a vote in ignorance or to abstain."

Combined Strategies (using Plurals)

Just keep in mind that approach number one and approach number two (using "he or she," and using "they") should not be used together. And if you combine two and three (using "they" and sticking with nouns), the nouns will need to be plural. Here's a combo that uses "they":

"When average American voters go to the polls on voting day, **they are** typically familiar only with the most highly publicized electoral races, and when [] confronted with choices between candidates for more obscure positions, such as nonpartisan judicial candidates, **the voters** must decide whether to cast **their votes** in ignorance or to abstain."

Completely Restructured Sentence

We've focused on substitutions here for the sake of space, but restructuring your entire sentence is an option as well. Here's just one example:

"Typically familiar only with the most highly publicized electoral races, the American voter who is confronted on election day with choices between candidates for more obscure positions, such as nonpartisan judicial candidates, must decide whether to vote in ignorance or to abstain."

There's nothing wrong with substitutions, but we hope you'll enjoy playing with your sentence structures at least some of the time. Not only will that make you a more flexible and creative writer, it's also a skill you'll need when paraphrasing work by others, as we'll explain in Chapter 7.

CHECKLIST FOR WRITING WITH STYLE

Do

✓ Use signposts and transitions to guide your reader through your paper.

✓ Hedge your claims to enhance your credibility.

✓ Present old information first to provide a context for what follows and new information last to emphasize it.

✓ Choose precise verbs to increase the clarity of your writing.

Don't

✓ Use grandiose claims and clichés.

✓ Use unnecessarily stuffy vocabulary and bloated sentences.

✓ Refer to everyone as "he."

SELECTING AND CITING SOURCES

In Chapter 1, we said the process of choosing sources and citing them effectively is more interesting than many people think. We hope our discussion in Chapter 3 on how to interact with sources when writing literature reviews gave you an idea of why this might be: choosing appropriate sources and writing about them in a way that makes your own contribution to the scholarly conversation clear are crucial to establishing your credibility.

In this chapter, we'll explain how to select appropriate primary and secondary sources (and what the difference between those is), when to quote directly and when to paraphrase, and how to avoid unintentional plagiarism. We'll also offer an overview of two of the citation styles commonly used in political science.

Defining Primary and Secondary Sources

Choosing good sources can be a complex business and is a skill you'll develop over time. In Chapter 1, we noted that beginning academic writers tend to think the purpose of using sources is to "look stuff up," or in other words, to accumulate information. But once you begin to think of yourself as a participant in the scholarly conversation we've described throughout this book, you'll start to see that while sources do sometimes serve primarily to provide facts or data, their more typical role is to give you a sense of what has been said in the conversation before you joined it so you can figure out what would count as a worthwhile contribution on your part. (You don't want to be the person who shows up late to the party and keeps telling jokes other people already told before you got there.) In a literature review, which we described in Chapter 3, your task is to provide an overview of the conversation thus far.

In many writing assignments, instructors may direct you to seek out either primary or secondary sources, or some of each. Of course, they may also simply tell you to use outside sources in your work without telling you which kind they think you need. When that happens, you can often figure out what kind of sources would be appropriate by thinking through the purpose of the assignment and what it is designed to help you learn.

First, we should make sure you know exactly what the difference between primary and secondary sources is. (We should also mention that not all disciplines talk about sources in the same way that political scientists do, so don't be surprised if what scholars call "primary" and "secondary" is different in history than in political science, and different again in biology.) In political science, primary sources are sources of information or data. They are the raw materials from which to build research. For instance, if you want to know how voters feel about a proposed tax law, survey data would be a good primary source. If you want to know how media outlets have been describing that same tax law, then newspaper articles, television news broadcasts, and tweets by journalists would all be relevant primary sources.

Primary sources = stuff you are analyzing

Other people's research and interpretations—their contributions to the scholarly conversation—are secondary sources. If you look directly at survey data about voters' views on the proposed tax law, you are looking at a primary source; if you look at somebody else's analysis of that survey data, you are looking at a secondary source. Sometimes, the same document could be either a primary or a secondary source, depending on your purpose for consulting it. For example, if you want to think about the role of gender in Marx's political theory because you're really interested in Marx, texts by Marx will be your primary sources, and feminist (and other) theorists' analyses of Marx's texts will be your secondary sources. You'll be participating in a scholarly conversation about Marx. However, if you want to figure out how feminist theorists' views of the role of gender in Marx's theory changed between 1970 and 1990, feminist analyses written during that period will become your primary texts, and you'll be participating in a scholarly conversation about this aspect of feminist theory.

Secondary sources = Other people's ideas about the stuff you are analyzing

This leads us back to choosing appropriate sources based on what a writing assignment is designed to help you learn. If you are expected to analyze data, whether that data is quantitative (survey responses about the proposed tax law) or qualitative (media rhetoric about the proposed tax law) or both, you'll need primary sources. If you are expected to participate in a scholarly conversation of any kind, you'll need to read secondary sources in order to know what others have been saying. For an assignment like that hypothetical one we described in Chapter 2 ("In a five- to ten-page paper, discuss Fukuyama's argument about the end of history. You may draw on outside sources to make your case, as long as you cite them properly. . . ."), you're explicitly being asked to participate in a conversation with Fukuyama, so secondary sources would be most appropriate. (Only two or three, as we noted in Chapter 3; you'd be looking for some sense of what the conversation has looked like so far, but you wouldn't have time to familiarize yourself with all of it or address many other people's ideas in so few pages.)

Locating Credible Primary Sources

Once you know which type(s) of sources you're looking for, you need to figure out where to find credible ones. Deciding what counts as credible depends in part on your research question; we'll start with primary sources first. Almost anything could conceivably be a primary source: if your question has to do with representations of politics on television or in video games, for example, then relevant television shows or video games will be your primary sources. If your question is about how lawsuits over some particular issue have been resolved in different counties or states, then court records might be what you need. And as we noted earlier, if you're studying the views of living people, those people will be your sources, and you'll collect your data from them via interviews, surveys, or both.

Often, however, what you'll need is data somebody else has already collected, and when that happens, you need a way to figure out whether that data is credible. Ideally, you should already know something about data collection and be able to evaluate whether the instruments used to collect the data were appropriate. (If a survey or interview was used, was it properly designed to elicit honest responses? Was the sample the data was collected from selected in a way that was not likely to introduce biases? Were key concepts operationalized in a way that makes sense?) But we realize that you may still be at an early stage in developing your research skills and may not be able to answer those questions with any confidence. In that case, you'll need to evaluate the credibility of the person (or people) who collected the data, and the credibility of the venue in which the data appears.

Academic researchers are usually credible sources, both because they have spent time learning how to collect accurate and ethical data, and because other experienced researchers are likely to look at the data and point out any problems they see in it. Thus, data that has been collected by scholars and that is housed on an official university website is usually a safe bet. Government websites at all levels (federal, state, county, or city) are also credible sources of data, as are sites that are maintained by well-recognized international organizations such as the United Nations and the World Bank. Data provided by foreign governments should in principle be credible, but it's important to consider the nature and reputation of the government in question and its purpose for making the data available. (If a dictatorship that is under international pressure because it has been accused of human rights violations suddenly releases data that highlights the well-being of its citizens, you'll want to inspect that data with a skeptical eye.) Similarly, whenever you want to use data collected by an organization of any kind, it will be important to consider that organization's reputation and its purpose for collecting the data and offering it to the public. So, for example, it's a good sign that the Global Terrorism Database (GTD) is published by and hosted at the University of Maryland.[1] If you find a similar database published by an unknown organization, a foreign country or institution, or even a known organization that is partisan (even one you might sympathize with), you should be more skeptical. Maryland's GTD might have errors, too, as might any piece of scholarship, but you can at least assume it exists out of sincere nonpartisan academic curiosity.

Academic researchers and governments are usually credible sources of data, but there are exceptions. Always consider why the data is being made available and why it is housed where it is.

Okay, we just said that academic researchers are usually credible, but they sometimes have political axes to grind like anyone else. If you find data that was collected by an academic but that is stored in an unexpected place, such as a personal website rather than a university site, it will be wise to think about why that is. On the one hand, it may be a matter of speed and convenience, and the data may be entirely legitimate; on the other, it may be a matter of flawed research practices or political bias, and the data may be unwelcome on a university site. When in doubt, ask someone for help.

The Librarian is Your Friend

These days, you can do most of your library research from the comfort of your dorm room, apartment, or the coffee shop. That's reasonable—professional scholars also like not having to leave their offices. But we encourage you to check out your library. There will be stuff there that's not online yet. But, more importantly, there are librarians! Librarians love helping people, especially with research questions, citation questions, and pretty much any other kinds of questions, such as helping them evaluate the quality of sources. Your college library might even have a dedicated government records librarian, or subject-area specialists not just for political science but Asian studies,

continued

physics, and classics. Librarians don't get into their business because they like to shush people. They get into it because they love knowledge and helping others make sense of it. Here's a case in point: at the University of Michigan, the political science subject librarian was the person who was able to point an honors student to the GTD we mentioned earlier, even though it's not a University of Michigan library resource. This information made the student's honors thesis possible.

Locating Credible Secondary Sources

The range of secondary sources that are likely to be appropriate is much smaller because what counts as a credible secondary source tends to fall into fairly narrow confines. When you are participating in a scholarly conversation, you are generally expected to cite academic sources: books published by academic presses and articles published in academic journals, all of which have been written by people who do research for a living. Sources directed toward a more general audience—newspaper or magazine articles, or blog posts, for example—might be relevant as well, if your topic is a particularly current one, but the majority of your sources should almost always be scholarly ones.

However, there is also a category of material that might be more difficult to evaluate. At the time of this writing, nearly all academic books and journals appear in print, and many of those also appear in digital formats such as e-books or online journal editions. But what about material written by professional political scientists that appears only online? In this case, you'll need to evaluate not only who has authored the work, but also who its target audience is. Does it appear to be written for scholars, or for general readers? Is it on a professional or institutional site, or on a personal one? Is it an article, a policy paper, or just a blog post? Does it cite sources, or only offer opinions?

Academic sources = books published by academic presses, peer-reviewed journals

General sources = newspapers, magazines, blogs, television and radio

This is a matter of understanding who the relevant participants in (and audience for) the conversation are. Think of it this way: if you were discussing football plays with a group of professional coaches, you probably wouldn't spend a lot of time referring to your next-door neighbor's views, no matter how well informed an armchair quarterback he or she might be. You'd stick to the experts: coaches, former coaches, and professional players. Sometimes, though, even coaches probably do talk to their neighbors, so while you can rule your neighbor out for purposes of this conversation (unless you happen to live next door to a professional coach!), you can't automatically rule the coach (or the scholar) in, until you're sure you know who the coach is talking to.

Reference Managers (or: Don't Waste Your Time by Typing Citations)

Reference managers are specialized database programs dedicated to scholarly work. They help you keep track of the materials you find, keep your reading notes organized, and, most importantly for our purposes, help you

continued

insert citations and bibliographies into your work with the simple push of a button. They do take a little bit of effort to learn and diligent use for them to remain useful, but they are extremely valuable time savers. Because library resources are now standardized into a digital format, you can copy citations (and sometimes entire texts) into your reference library with a single click. And you can format citations and bibliographies to the style you want to use by selecting from a menu, instead of learning a new set of complicated rules.

There are numerous different reference managers available, some commercial and costly, some free. So do yourself a favor and check out RefWorks (free to some university students), EndNote (commercial), Zotero (free), Mendeley (free), and BibTeX (free).

CHECKLIST FOR CHOOSING CREDIBLE SOURCES

Do

✓ Know the difference between primary and secondary sources.
✓ Consider who produced the source and why.
✓ Ask a research librarian for help.

Don't

✓ Use a non-scholarly secondary source in a scholarly conversation, unless you have a very good reason.

Paraphrasing Versus Quoting Versus Summarizing

When you participate in a scholarly conversation, you need to include others' ideas in your paper so your readers will know what you are responding to. There are three ways to convey other writers' views: summarizing them, paraphrasing them, and quoting them directly (Table 7.1).

TABLE 7.1 **The Differences Between Summary, Paraphrase, and Direct Quotation**

Summary	• A compressed version of someone else's idea or argument • Shorter than the original
Paraphrase	• A restatement of someone else's idea in your own words • May be shorter, longer, or the same length as the original
Direct quotation	• Someone else's exact words, set off with quotation marks

Every subfield of political science includes summaries, but readers have subfield-specific expectations about when paraphrasing and quoting are appropriate. (And here we should note that paraphrases, summaries, and direct quotations all need to be cited, whether through footnotes or parenthetical citations. We'll get back to how to cite correctly later in the chapter.)

In quantitative-formal work, it's generally important to report on what other researchers have said, but how they said it—the exact words they used—doesn't matter. In interpretive work, where textual and conceptual analysis is involved, minor differences in word choice can change the meaning of an interpretation, so it's important to show exactly what another writer said and how that writer said it.

In Chapter 1, when we talked about subfields and genres, we said that readers have different expectations for what we grouped into the quantitative-formal and the qualitative-interpretive categories of research. As a quick rule of thumb, you can assume that direct quotations are not appropriate (except under the rarest of circumstances) in quantitative-formal genres. In contrast, they are a fundamental unit of evidence in interpretive genres, and in some qualitative ones that collect evidence via interviews or participant observation. Thus, *direct quotations are always necessary in political theory*, where textual analysis is always involved. *They are sometimes necessary in public law*—when analyzing court cases, for example. And *they may be necessary in American politics or comparative politics* when analyzing speeches or propaganda. In contrast, direct quotations will usually be inappropriate in quantitative analyses of voter behavior, economic policies, or correlates of war. *If you are writing a paper using the IMRD (Introduction, Methods, Results, Discussion) structure, you will generally not want to include direct quotations.* Instead, summarize and paraphrase sources.

Even when writing in a genre in which direct quotations are expected, you'll need to paraphrase some of the material you use from other sources. Otherwise, your paper will look like one long string of quotation marks and your own voice will disappear. In those cases, deciding what to quote directly and what to paraphrase is a matter of deciding which passages include key concepts and wording. If nothing important will be lost when you paraphrase, then you should paraphrase.

The mechanics of direct quotation are fairly straightforward and probably already familiar, but we often see writers get some of the details wrong, so we'd like to offer a quick look at how to integrate quoted material into your own writing. Then we'll turn to paraphrasing, which is a bit more complicated.

Integrating Direct Quotations into Your Writing

While we're all familiar with how direct quotations look on the page, fiction readers sometimes make the mistake of formatting them as if they were passages of dialogue instead of framing them in ways that help make it clear what is being quoted and why. It is almost never—in fact, we'll go so far as to say really, truly never—appropriate to simply plop a sentence from another writer's text into your paragraph and surround it with quotation marks. We'll give some examples of how to quote properly here, and we'll explain how and when to include footnotes and parenthetical citations in the final section of the chapter.

This passage, from a senior honors thesis, includes a well-integrated direct quotation:

> MacKinnon stressed that a law cannot be separated from the society in which it exists. She states, "In the context of international humanitarian law, to look to coercion to define rape is to look to the surrounding collective realities of group membership and political forces, alignments, stratifications, and clashes."[2]

The first sentence makes a claim about what MacKinnon says, one that is, in effect, an interpretive summing-up of the direct quotation that follows. The mistake many inexperienced writers make with a passage like this one is to leave out the phrase "she states" and jump right into the direct quotation. In that case, the passage would look like this:

> MacKinnon stressed that a law cannot be separated from the society in which it exists. "In the context of international humanitarian law, to look to coercion to define rape

is to look to the surrounding collective realities of group membership and political forces, alignments, stratifications, and clashes."

It may seem obvious in this version that the direct quotation has been taken from MacKinnon, but this "dropped-in" quotation creates awkward writing. It is the writer's job to attribute all direct quotations to their original writers by building a bridge between the original writer and the quoted words into the sentence in which those words appear (Table 7.2).

TABLE 7.2   **Integrating Quotations Into Your Text**

Well-Integrated Quotation	Poorly Integrated Quotation
MacKinnon stressed that a law cannot be separated from the society in which it exists. **She states,** "In the context of international humanitarian law, to look to coercion to define rape is to look to the surrounding collective realities of group membership and political forces, alignments, stratifications, and clashes."	MacKinnon stressed that a law cannot be separated from the society in which it exists. "In the context of international humanitarian law, to look to coercion to define rape is to look to the surrounding collective realities of group membership and political forces, alignments, stratifications, and clashes."

There are many ways to integrate others' words into your own sentences and paragraphs. One alternative to the approach above would be to substitute a colon for the phrase "she states":

MacKinnon stressed that a law cannot be separated from the society in which it exists: "In the context of international humanitarian law, to look to coercion to define rape is to look to the surrounding collective realities of group membership and political forces, alignments, stratifications, and clashes."

The colon makes this work because one function of the colon is to assert a relationship of equivalence. Thus, it tells the reader that what follows the colon contains the same meaning as what precedes it.

Obviously, you can't rely only on "she states" or on colons, or your prose will become monotonous. It's also important to pay attention to how the verbs you choose to introduce another writer's words signal your own stance toward that writer's ideas, as we noted when discussing precise verbs in Chapter 6. So you'll want to choose carefully between introductory phrases such as "According to MacKinnon," "MacKinnon argues," and so on. Consider the small but crucial difference between these two options in Table 7.3.

TABLE 7.3 The Nuances between Phrase Choices

"According to MacKinnon . . ."	"I'm just telling you what MacKinnon said. I'm not taking a position on it right now."
"As MacKinnon notes . . ."	"MacKinnon said this true thing that I am passing on to you."

You'll also want to get comfortable integrating parts of other writers' sentences into your own. In the example we've been working with, the student could have chosen a more abbreviated approach that would look something like this:

MacKinnon stressed that a law cannot be separated from the society, or the "collective realities of group membership and political forces, alignments, stratifications, and clashes," in which it exists.

As you can see, you have many choices for talking about others' ideas well. But you can also put things in ways that are confusing, awkward, or misleading, so be vigilant.

Editing Quoted Text for Grammatical Consistency

Sometimes, in order to make a portion of someone else's text grammatically consistent with your own, you'll need to edit it. The tools you can use to do this are **brackets** ([ ]) and **ellipses** (...). Brackets signal that you have changed or inserted a word or phrase for the sake of clarity and grammatical correctness. For example, if you quote a writer whose text is in first person, but you are talking about her in third person, you can change the word "my" to "her," as long as you put "her" in brackets, and if you are writing in past tense while the writer you're quoting wrote in present tense, you can adjust that, too:

> Anderson noted that she hadn't decided how to vote yet because "[her] research on key elements of the policy questions [was] still incomplete."

If what Anderson really wrote was something longer, such as "My mind is not yet made up because my research on key elements of the policy questions regarding leash laws and public parks is still incomplete," but your reader already knows the issue in question has to do with leash laws, you can shorten the passage with an ellipsis:

> Anderson noted that she hadn't decided how to vote yet because "[her] research on key elements of the policy questions ... [was] still incomplete."

Brackets signal that you have changed or inserted a word or phrase for the sake of clarity and grammatical correctness.

An ellipsis indicates that you have removed material for the sake of brevity.

Sometimes you may need an ellipsis at the end of a sentence you choose to shorten. If Anderson had written "My mind is not yet made up because my research on key elements of the policy questions regarding leash laws and public parks is still incomplete and is likely to remain incomplete for the next several weeks," your version would look like this:

> Anderson noted that she hadn't decided how to vote yet because "[her] research on key elements of the policy questions . . . [was] still incomplete. . . ."

Take note that an ellipsis that appears at the end of a sentence includes a fourth dot that serves as a period.

Don't use an ellipsis when you leave out words at the beginning of a passage you are quoting. As you can see in our examples above about Anderson's voting plans, we have left out her first few words—actually, we have paraphrased them—and have used quotation marks to indicate where the direct quotation begins.

We should also mention that when you quote a very long passage—five lines or more, if you are using Chicago Style, for example—it should be formatted as a block quote that is set off into a "block" of its own, indented, and not framed by quotation marks. A block quote looks like this:

> It needs to be introduced by a phrase of your own, just like quoted material that appears within a standard paragraph. (You may notice that we've been using a lot of colons; this is a fairly common approach.) It may include several sentences, and like other direct quotations, it may include ellipses if you need to leave . . . material out. You will want to use block quotes sparingly; a paper that includes too many of them suggests that its writer either has difficulty making

judgments about what really needs to be quoted or is too lazy to figure out how to summarize, paraphrase, and develop his or her own ideas. When you do use them, keep them as short as possible. (LaVaque-Manty and LaVaque-Manty 2015)

> When using a block quote, exclude quotation marks but include an in-text citation, as in the example directly above.

It's also generally frowned upon to end a section or even a paragraph with a block quote. The content of a long portion of quoted text requires some interpreting or explaining, and the explanation should follow the long quotation. Here's an example from that senior thesis:

Catharine MacKinnon elegantly explains the impact that laws against rape have on victims and communities:

> The degree to which the actualities of raping and being raped are embodied in law tilt ease of proof to one side or the other and contribute to determining outcomes, which in turn affect the landscape of expectations, emotions, and rituals in sexual relations, both every day and in situations of recognized group conflict. (940)

Thus, people who have never been raped, and perhaps never will be, are still unconsciously impacted by the rape legislation in effect in their jurisdiction. The outcomes of rape cases tell society what is acceptable and what is not, and dictate how people expect to treated by others. Defining the crime of rape impacts how rape is understood by the populace in addition to aiding the prosecution of the crime which is so destructive to individuals and communities.

The writer explains the passage in language the reader will be sure to understand, translating "affect the landscape of expectations, emotions, and rituals in sexual relations" into "tell society what is acceptable and what is not, and dictate how people expect to be treated by others."

Shorter passages that don't require block quotes should generally be interpreted for readers, too. After all, quoting that passage won't help your argument if your reader doesn't understand what it means, or why you think it's important.

CHECKLIST FOR DIRECT QUOTATIONS

✓ Did you transcribe the quoted passage accurately?
✓ Did you set it off with quotation marks?
✓ Did you build a bridge between the quotation and your own words?
✓ If the quoted passage takes up more than four lines in your paper, did you create a block quote?
✓ Did you interpret quoted passages for your reader?
✓ Did you cite the source you quoted in the text or in a footnote?

The Art of Summarizing

To summarize means to compress. Imagine that you are writing a response paper about *The Evolution of Cooperation*, by Robert Axelrod, and you need your reader to know what the book is about so you can respond to it. *The Evolution of Cooperation* is about 200 pages long, but you are writing a response paper with a 300-word limit so you need to **summarize** it briefly, perhaps in one sentence: "*The Evolution of Cooperation* uses computer modeling to demonstrate that self-interested actors can be expected to cooperate when they know they will have to continue to interact over a long period of time." Summaries can be longer than this, of course, depending on how much space you have available, but their role is to convey a work's most important argument or findings ("self-interested actors can be expected to cooperate when they know they will have to continue to interact over a long period of time") and the methods used to arrive at those findings ("uses computer modeling").

Summarizing can take place at any point in a paper—whenever you need to introduce a writer into your scholarly conversation—but much of it occurs in introductions, and particularly in literature reviews, where many books and articles need to be addressed in a short space. Here's an example taken from the literature review in an undergraduate research proposal:

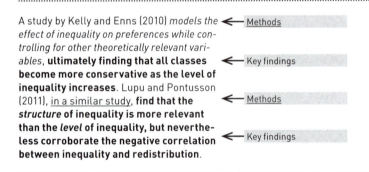

A study by Kelly and Enns (2010) *models the* ← Methods *effect of inequality on preferences while controlling for other theoretically relevant variables*, **ultimately finding that all classes** ← Key findings **become more conservative as the level of inequality increases**. Lupu and Pontusson (2011), <u>in a similar study</u>, **find that the** ← Methods **structure of inequality is more relevant than the *level* of inequality, but nevertheless corroborate the negative correlation** ← Key findings **between inequality and redistribution**.

Here, the writer efficiently characterizes two prior studies in terms of both what they found and how they found it, using only two sentences. Such compression allows her to address a large amount of literature in only a few pages in her own paper.

The Art of Paraphrasing

The purpose of paraphrasing is to allow you to convey the ideas contained in a small portion of someone else's text, such as a sentence or two, while also maintaining your own voice. To paraphrase something means to put it in your own words. Many writers are surprised to learn that it also means putting it into your own syntax, or sentence structure. The need to change syntax is what makes paraphrasing complicated; it's not enough to paste somebody else's sentence into your paper and replace its vocabulary, you also have to change the way the sentence is built, but without changing the meaning of the ideas involved.

Paraphrasing = changing vocabulary + changing syntax without changing meaning

As we noted earlier, there are genres in which direct quotations either aren't acceptable or must be used extremely sparingly—in quantitative empirical papers, for example. In those genres, you will need to do quite a bit of paraphrasing.

Let's figure out how to paraphrase this: "The quick brown fox jumped over the lazy dog." Many writers would paraphrase this sentence by changing only the words that are easiest to change and ignoring the syntax, resulting in something like this: "The speedy russet fox leaped over the indolent canine." There are two problems with this sentence. First, it probably doesn't sound like the writer usually sounds. (Who says "indolent canine"?) Second, despite its weirdness, this sentence remains insufficiently transformed; in fact, the syntax is exactly the same as it was in the original.

At this point, you might be wondering what would count as good substitutes for "fox" and "dog." The answer might be that there aren't good substitutes. It depends on the context. If this sentence were part of a longer story, we might know what kind of dog this was, and it might be fair to say "hound," or "beagle." "Fox," on the other hand, might need to remain "fox."

And what about "lazy"? "Idle" might do. It's definitely better than "indolent." This issue becomes more important when you're dealing with academic language, which often includes technical terms that can't be changed without losing the meaning of the ideas they're meant to convey. In political science, this might include words such as "voter," "survey," "data," "policy," and "sovereignty."

The Thesaurus is Often Not Your Friend

When you want to paraphrase or simply keep your prose from getting monotonous, you might turn to a thesaurus. There's one available behind a single keyboard shortcut, after all! But because academic vocabularies are specialized, even when they *seem* to use everyday words, choosing anything a thesaurus picks correctly is tricky. If you know how to choose correctly, you didn't need the thesaurus; if you don't, you risk losing credibility. Besides, variety in your word choices, especially in the important ones, is less important than consistency and precision.

One strategy that might help you develop an appropriate specialized political science vocabulary is to pay attention to the vocabulary used in the texts you read for your courses and keep notes on which terms show up often.

Let's look at another example and consider the issue of syntax. This time, we'll work with the kind of statement you might find in a published research article: "Political knowledge is a central concept in the study of public opinion and political behavior."[3] (Again, we've chosen a short sentence because those are the most difficult to work with.) Changing syntax requires changing word order, by which we mean changing the arrangement of phrases and clauses. That will be particularly important here, because our sentence contains three technical terms for which we can't offer substitutes: "political knowledge," "public opinion," and "political behavior." (That's six out of fifteen words!)

You might notice that our example contains only one clause. That makes rearranging things a bit more difficult. Still, it can be done. Here are a couple of possibilities:

1. When doing research on public opinion and political behavior, political knowledge must be taken into account.
2. The fields of public opinion and political behavior cannot avoid addressing issues of political knowledge.

As you have probably guessed, paraphrasing well is time consuming; it takes careful thought to rearrange syntax and replace vocabulary without changing the meaning of the original. However, paraphrasing is also an important skill to master in order to avoid plagiarism. We offer more advice about avoiding plagiarism in the following section.

CHECKLIST FOR PARAPHRASING

✓ Are there key words that can't be changed without loss of meaning?
✓ Did you change the words that can be changed?
✓ Did you transform the structure of the sentence or sentences you are paraphrasing?
✓ Does the final product sound the way you sound in the rest of your paper?
✓ Did you cite the source you paraphrased?

Understanding and Avoiding Plagiarism

Plagiarism is taking the words or ideas of others and passing them off as your own. There are many varieties, ranging from buying or copying entire papers to copying just a few words. You might think of plagiarism as a combination of lying and stealing; when you plagiarize, you pretend an idea is yours when it is not, and you steal the credit for that idea from the writer it actually belongs to. Plagiarism is an ethics violation that can result in failing a paper, failing a class, or getting suspended or expelled from a university. We prefer not to think about it in terms of crime and punishment, however, but in terms of skill and credibility. While there are certainly writers who plagiarize intentionally, we have also encountered a large number who simply didn't understand how to use their sources well. If you're reading this book, we assume you want to write awesome, original papers, not cheat your way to an A.

One thing many less experienced writers have difficulty figuring out is what actually needs to be cited. For example, lots of information that you first encounter in another writer's text is actually what we call "common knowledge" for most political scientists, or at least most political scientists in a particular subfield, and common knowledge doesn't need to be cited. The fact that smaller samples tend to produce more bias than large ones is common knowledge, but that might not be obvious to someone who only learned about the concept of representative sampling last week. The good news is that what counts as common knowledge becomes increasingly clear as you spend more time in a discipline; the bad news is that there isn't any way to speed this process up, aside from taking more classes and reading more work in your field.

An Example of Local Common Knowledge

The fact that statistical analysis is more reliable when applied to large samples than it is when applied to small ones is common knowledge among political scientists who do quantitative research. Small sample size is often referred to as "the small-n problem," where "n" means "number of cases in the sample." A student taking his or her first research methods class and encountering this information for the first time might not know this, and might assume it would be necessary to cite a source to support the claim that it's better to use a large sample when possible.

We have to admit that citing ideas that don't need to be cited isn't good for your credibility: it signals that you are a newbie. However, not citing things that do need to be cited is worse. Instructors have more patience with beginners than they do with cheaters, and citing too much won't get you expelled. So, when in doubt, include a citation.

Citing things you don't need to cite shows that you are inexperienced and tells readers that you don't know what counts as "common knowledge" in the field you're writing in. But failing to cite something that should be cited is plagiarism, even if you didn't intend to cheat. So when in doubt, err on the side of citing too much, rather than too little.

We're pretty sure you know that citations must always be provided for direct quotations, and it's difficult to summarize an entire article or book without saying who wrote what you're summarizing, so that's a no-brainer, too. Which brings us back to paraphrases. **Paraphrases must always be cited**, and paraphrases must always thoroughly paraphrase the original work, where "thoroughly" means changing both vocabulary and syntax, as we described earlier.

Patchwork Plagiarism

One common form of plagiarism is called patchwork plagiarism, where writers stitch together pieces of sentences written by others into a sort of crazy quilt that they then submit as a paper. Most writers who do this don't realize that they are plagiarizing. They think replicating parts of sentences, as opposed to entire sentences, is okay. But not only is this plagiarism, it also produces an inconsistent voice that often makes the plagiarism easy to spot. If you take phrases here and there from five different writers, it is unlikely that those phrases sound like you sound when you write from scratch.

Patchwork plagiarism tends to result from thinking about your sources as places where you look stuff up rather than as lines of thought generated by real people whom you can converse with. If you're keeping track of a scholarly conversation, you'll want to remember who said what. If you're looking stuff up, you'll just write that stuff down and regurgitate it, rudely, as if it's your own idea, which can be embarrassing when you realize that the person reading your paper has a reasonably good chance of knowing who actually said it first.

To avoid this type of plagiarism, we recommend that you generate a careful method of note-taking in which you track where every idea you've collected came from, and that you get used to spending serious amounts of time paraphrasing other people's words. Develop a system that makes it clear whether what you've put in your notes is a direct quote or is in your own words, by highlighting any direct quotes in yellow, for example. This can help you avoid the mistake of including un-paraphrased (or insufficiently paraphrased) ideas in your papers because you've forgotten the words you're transcribing weren't originally your own. Include not only authors and titles when you take notes, but also page numbers, so you can find key passages again. You may find all of this easiest to do with a reference manager, as we noted earlier in this chapter.

In the next section, we'll walk you through the process of citing other people's work in your text, both in the body of your paper and in your bibliography, in Chicago Style and American Political Science Association (APSA) Style.

Common Citation Styles in Political Science

Chicago Style and APSA Style are citation systems commonly used in political science. In this section, we'll describe when and how to cite your sources, and we'll provide examples of how to do so using each of these styles. Citation styles change over time, which makes it important to be sure you're using the most recent edition of your required style guide. In what follows, we are using the 16th edition of the *Chicago Manual of Style* and the *Style Manual for Political Science* from 2006.

Why to Cite
As we hope is clear from the rest of this chapter, citing sources is necessary to avoid plagiarism—but it also does much more than that:

- It allows you to give credit where credit is due—to the people whose ideas you're using.
- It provides a research guide for others. Just as we recommend looking carefully at what gets cited in the works you read to get a sense of the conversation that has preceded you, your own citations provide that same sort of resource for others.

- It helps establish your credibility by showing that you've read the sources that are relevant to your conversation.
- It lets you signal your own contributions. Those ideas not attributed to others through your citations are the ideas that originate with you.

When to Cite

In the previous section, we explained what needs to be cited, but we probably didn't answer many of your questions about where the citations in the body of your paper should actually appear. One purpose of citing your sources is to let readers know (immediately) where they can find those sources themselves. When you summarize, paraphrase, or directly quote another writer's work, you need to include a parenthetical citation (author year) or footnote[4] at the point where you first use any portion of that work. In Chicago Style, you may use either footnotes or parenthetical citations. (We'll provide examples of each below.) Your instructor may prefer one or the other; if it's up to you to choose, be consistent in your approach—use footnotes or parenthetical references, but not a mishmash of both. In APSA Style, parenthetical references are always used to cite sources; footnotes are reserved for explanatory material that doesn't fit in the body of the paper itself, and you should minimize your use of them.

So, where do these citations go? They should immediately follow each direct quote or paraphrased passage. That means that you may need to include several footnotes or parenthetical

In **Chicago Style**, you can either use footnotes or use parenthetical citations + a reference list.

In **APSA Style**, footnotes are reserved for information and comments; you must use parenthetical citations + a reference list to cite your sources.

references in a single paragraph. If you are summarizing an article or book, you may be tempted to place your citation at the end of a paragraph or sequence of paragraphs. However, this is similar to reporting what one friend told you to a second friend without telling the second friend when and where the original conversation took place until after you've finished your whole long story. ("So, when was this?" your friend asks when you've finally stopped talking. "Where were you guys, exactly?")

> To prevent confusion, it's often simplest to name your author and text directly in the first sentence of your summary and include a reference either within that sentence or at the end of it. Here's a version that works for both Chicago Style and APSA Style: "In *The End of History and the Last Man* (1992), Francis Fukuyama argues . . ."

Once you've made it clear what source you are summarizing, you don't necessarily need to include additional parenthetical references, unless you directly quote or paraphrase a key idea. At that point, another citation must directly follow the quoted or paraphrased passage. When you include a direct quotation, you should also include the number of the page(s) on which the quoted passage appears.

Footnotes, used only in Chicago Style, include complete bibliographic information the first time a text is cited.

Here's a footnote for a book about footnotes.[5]

Here's a version of the same footnote that includes a page number for a direct quotation.[6]

---

5. Jane Smith. *Fake Text Title for Footnote Sample.* (New York: Excellent Publisher, 1998).

6. Jane Smith. *Fake Text Title for Footnote Sample.* (New York: Excellent Publisher, 1998), 14.

If you need to cite the same page twice in a row, your footnote will say "Ibid.," which means "in the same place."[7] If you cite the same source, but refer to a different page, you can still use "Ibid.," but you'll add the new page number.[8] If you cited a different source after Smith,[9] you would not be able to use "Ibid." the next time you cited Smith, but you would use a shortened version of the Smith footnote.[10]

CHECKLIST FOR WHEN TO CITE

✓ Did you cite each source the first time you used it?
✓ Did you cite each time you paraphrased?
✓ Did you cite all direct quotations?

How to Cite

At this point, we'd like to provide examples of parenthetical citations, footnotes, and reference list formatting for a variety of sources. However, we're offering these only to give you a sense of the similarities and differences between these two styles. While it may be easy to memorize the rules for creating parenthetical citations, we don't recommend that you try to memorize the rules for constructing footnotes and bibliographic references. There are simply too many details to keep track of, especially if you're also using different citation styles for papers in other courses. We ourselves use a reference manager, as we recommended doing earlier, and we check its output carefully against citation guidelines and sample references.

If you don't find examples of the materials you cite in what we've written below, online guides are available for both APSA Style and Chicago Style.[11] If you don't find what you need there, either, we recommend consulting a librarian.

Parenthetical Citations

Parenthetical references are built the same way in both APSA Style and Chicago Style (Table 7.4). A reference that directs the reader to the overall text under discussion requires an author's last name, followed by a date, with no punctuation appearing between these two items. That looks like this: (Smith 1998). A reference that directs the reader to a specific page on which a directly quoted passage or particularly important paraphrased idea appears should include a page number after the date, with a comma placed between the year and the page number: (Smith 1998, 135). The purpose of such in-text references is to direct readers to a specific source in the reference list that appears at the end of the paper. Without the reference list, in-text citations wouldn't be very helpful. And, we should note, every parenthetical reference that appears in your paper must correspond to a reference that is listed in your reference list, and vice versa; no source should be listed in your reference list if you didn't cite it in the body of your paper.

TABLE 7.4 **In-text Parenthetical Citation Guide**

	Chicago Style Template	Chicago Style Example	APSA Style Template	APSA Style Example
Book or journal article with single author	(Author year, page #)	(Axelrod 1984, 17)	(Author year, page #)	(Axelrod 1984, 17)
Book or journal article with two or three authors	(Authors year, page #)	(Hardt and Negri 2001, 8)	(Authors year, page #)	(Hardt and Negri 2001, 8)
Book or journal article with four or more authors	(First author et al. year, page #)	(Barabas et al. 2014, 842)	(First author et al. year, page #)	(Barabas et al. 2014, 842)
Edited book	(Editor year, page #)	(Morgan 2011, 42)	(Editor year, page #)	(Morgan 2011, 42)

continued

	Chicago Style Template	Chicago Style Example	APSA Style Template	APSA Style Example
Chapter in an edited book	(Author year, page #)	(Hay 2013, 289)	(Author year, page #)	(Hay 2013, 289)
Website	(Site Owner year)	(APSA 2014)	(Site Owner year)	(APSA 2014)
Newspaper article	(Author(s) year, page)	(Romero and Neuman 2014, 1)	None	None

Newspapers do not get cited in text in APSA Style, only in footnotes.

References in a Reference List

Table 7.5 gives examples of how to style references in Chicago Style and APSA Style.

TABLE 7.5   **Reference List Guide**

	Chicago Style Template	Chicago Style Example	APSA Style Template	APSA Style Example
Book with single author	Author. year. *Title*. City: Publisher.	Axelrod, Robert. 1984. *The Evolution of Cooperation*. New York: Basic Books.	Author. year. *Title*. City: Publisher.	Axelrod, Robert. 1984. *The Evolution of Coopera-tion*. New York: Basic Books.
Book with two or three authors	Authors. year. *Title*. City: Publisher.	Hardt, Michael and Antonio Negri. 2001. *Empire*. Cambridge, MA: Harvard University Press.	Authors. year. *Title*. City: Publisher.	Hardt, Michael and Antonio Negri. 2001. *Empire*. Cambridge, MA: Harvard University Press.

	Chicago Style Template	Chicago Style Example	APSA Style Template	APSA Style Example
Edited book	Editor, ed. Year. *Title*. City: Publisher.	Morgan, Michael, ed. 2011. *Classics of Moral and Political Theory*. Cambridge, MA: Hackett.	Editor, ed. Year. *Title*. City: Publisher.	Morgan, Michael, ed. 2011. *Classics of Moral and Political Theory*. Cambridge, MA: Hackett.
Chapter in an edited book	Author. year. "Chapter Title." In *Title*, edited by Editor, page range. City: Publisher.	Hay, Colin. 2013. "International Relations Theory and Globalization." In *International Relations Theories*, edited by Tim Dunne, Milja Kurki, and Steve Smith, 287–305. Oxford: Oxford University Press.	Author. year. "Chapter Title." In *Title*, ed. Editor. City: Publisher, page range.	Hay, Colin. 2013. "International Relations Theory and Globalization." In *International Relations Theories*, eds. Tim Dunne, Milja Kurki, and Steve Smith. Oxford: Oxford University Press, 287–305.
Journal article	Author. year. "Article Title." *Journal Title* volume #, no. issue #: page range.	Saward, Michael. 2014. "Shape-Shifting Representation." *American Political Science Review* 108, no. 4: 723–736.	Author. year. "Article Title." *Journal Title* volume #: page range.	Saward, Michael. 2014. "Shape-Shifting Representation." *American Political Science Review* 108: 723–736.

continued

	Chicago Style Template	Chicago Style Example	APSA Style Template	APSA Style Example
Journal article with four or more authors	Authors. year. "Article Title." *Journal Title* volume #, no. issue #: page range.	Barabas, Jason, Jennifer Jerit, William Pollock and Carlisle Rainey. 2014. "The Question(s) of *Political Knowledge*." *American Political Science Review* 108, no. 4: 840–855.	Authors. year. "Article Title." *Journal Title* volume #: page range.	Barabas, Jason, Jennifer Jerit, William Pollock and Carlisle Rainey. 2014. "The Question(s) of *Political Knowledge*." *American Political Science Review* 108: 840–855.
Website	Site owner. Year. "Page Title." Access date. URL.	American Political Science Association. 2014. "For Students." December 12. http://www.apsanet.org/students.	Site owner or author. Year. "Page Title." URL (Access date).	American Political Science Association. 2014. "For Students." http://www.apsanet.org/students (December 12, 2014).
Newspaper	Author. "Title." *Newspaper Title*, date.	Romero, Simon, and William Neuman. "Cuba Thaw Lets Latin America Warm to Washington." *New York Times*, December 18, 2014.	None	None

Newspapers do not get cited in text in APSA Style, only in footnotes.

Footnotes

As we mentioned earlier, you can use footnotes instead of in-text citations + reference lists to cite sources in Chicago Style. In APSA Style, however, footnotes should only contain comments and substantive information. The one exception to this rule in APSA Style is newspaper articles, which must be cited in footnotes and may not be cited parenthetically or appear in your reference list (Table 7.6).

> Note that in footnotes, authors' first names always come first, while in the reference list, the last name comes first for single authors, and for the first author when a source has multiple authors.

TABLE 7.6 **Footnote Citation Guide**

	Chicago Style Template	**Chicago Style Example**	**APSA Style Template**	**APSA Style Example**
Book with single author	Author, *Title* (City: Publisher, year), page #.	Robert Axelrod, *The Evolution of Cooperation* (New York: Basic Books, 1984), 17.	None	None
Book with two or three authors	Authors, *Title* (City: Publisher, year), page #.	Michael Hardt and Antonio Negri, *Empire* (Cambridge, MA: Harvard University Press, 2001), 8.	None	None

	Chicago Style Template	Chicago Style Example	APSA Style Template	APSA Style Example
Edited book	Editor, ed., *Title* (City: Publisher, year), page #.	Michael Morgan, ed., *Classics of Moral and Political Theory* (Cambridge, MA: Hackett, 2011), 42.	None	None
Chapter in an edited book	Author, "Chapter Title," in *Title*, ed. Editor (City: Publisher, year), page #.	Colin Hay, "International Relations Theory and Globalization," in *International Relations Theories*, eds. Tim Dunne, Milja Kurki, and Steve Smith (Oxford: Oxford University Press, 2013), 289.	None	None
Journal article	Author, "Article Title," *Journal Title* volume #, no. issue # (year): page #.	Michael Saward, "Shape-Shifting Representation" *American Political Science Review* 108, no. 4 (2014): 723–736.	None	None
Journal article with four or more authors	First author et al., "Article Title," *Journal Title* volume #, no. issue # (year): page #.	Jason Barabas et al., "The Questions of Political Knowledge," *American Political Science Review* 108 , no. 4 (2014): 842.	None	None
Website	"Page Title," accessed date, URL.	"For Students," accessed December 12, 2014, http://www.apsanet.org/students.	None	None

	Chicago Style Template	Chicago Style Example	APSA Style Template	APSA Style Example
Newspaper article	Author, "Title," Newspaper Title, Date, page # or access date.	Simon Romero and William Neuman, "Cuba Thaw Lets Latin America Warm to Washington," *New York Times*, December 18, 2014, accessed December 18, 2014.	Author, "Title," *Newspaper Title*, Date, page #.	Simon Romero and William Neuman, "Cuba Thaw Lets Latin America Warm to Washington," New York Times, 18 December, 2014.

Unlike most sources, newspapers should be cited in footnotes rather than in text or in reference lists in APSA Style.

Note the difference between how the date is cited in Chicago Style footnotes (month, day, year) vs. APSA Style footnotes (day month, year).

COMMON ERRORS TO AVOID IN CHICAGO STYLE

✓ For footnotes and in-text citations with four or more authors, only list the first author's last name, followed by "et al." (Barabas et al. 2014).

✓ For reference list citations with more than one author, remember not to list all authors' last names first—only the first author's last name goes first (Barabas, Jason, Jennifer Jerit, William Pollock and Carlisle Rainey).

✓ For footnoted citations with more than one author, list the first author's first name first, followed by "et al." (Jason Barabas et al.,

"The Questions of Political Knowledge," *American Political Science Review* 108 (2014): 842).

✓ For reference list citations of websites, include the date you accessed the site (American Political Science Association. 2014. "For Students." December 12. http://www.apsanet.org/students.).

COMMON ERRORS TO AVOID IN APSA STYLE

✓ For in-text citations with four or more authors, only list the first author's last name, followed by "et al." (Barabas et al. 2014).

✓ For reference list citations with more than one author, remember not to list all authors' last names first—only the first author's last name goes first (Barabas, Jason, Jennifer Jerit, William Pollock and Carlisle Rainey).

✓ For reference list citations of websites, include the date you accessed the site (American Political Science Association. 2014. "For Students." http://www.apsanet.org/students (December 12, 2014).).

✓ Do not cite newspapers in your reference list.

✓ Do not cite anything but newspapers in your footnotes.

SEEKING AND USING FEEDBACK

Even on assignments that don't explicitly ask you to submit drafts and revisions, **seeking feedback is a natural—not remedial—scholarly habit**. We know that many students imagine that writing is a fast, easy process for "good writers," and our culture has long circulated an unhelpful image of "the writer" as a solitary genius working alone in a turret or cellar or hut. Perhaps such geniuses exist, but we've never met any. The writers we know ask colleagues they respect for careful, detailed feedback on their drafts. After all, the only way to know if you are making your ideas clear and meeting your readers' expectations is to have somebody (possibly several somebodies) read your work and tell you what made sense and what didn't.

Getting others to read your work is so important that there is an automatic feedback system built into the standard publishing process: any article submitted to a high-quality academic journal or book manuscript submitted to a respectable academic press will always be sent to other scholars for feedback before an editor decides whether it should be published. Much of the time, the writer will be asked to revise the work based on the feedback received and submit it again before the editor finally decides whether to accept it. And even after the editor says yes, the writer will usually be expected to make additional changes.

It's a good idea to get used to seeking and using feedback as early in your academic life as possible. We encourage you to begin building ways of getting feedback into your writing habits now. There are several ways you can do this:

1. **Meet with your instructor.** We know some instructors don't have time to meet with students about writing assignments, but many are happy to give input on proposals, outlines, or drafts. If your instructor is willing to give you feedback along the way, take advantage of that opportunity. (One of the reasons some instructors avoid commenting on drafts is that they have encountered this unhappy pattern: an instructor reads a student draft and encourages student to work on X, student works on X, turns in paper, receives a B+, and is unhappy because the student thought he or she did "everything that was necessary for an A." But the instructor was trying to help the student focus on the kind of changes that would improve the paper the most, not offering a magic formula for a good grade. So please remember that unless instructors explicitly grade your drafts, their feedback on anything before the final version isn't "pregrading.")

2. **Go to the writing center**, if your college or university has one. Writing centers exist to give students help with writing at all stages of a writing project, from conception to completion. If you receive a confusing assignment prompt, a tutor at the writing center may be able to help you decode it. Tutors can also help you make sure your paper is well organized and well supported, as well as give you tips on writing style.

3. **Enlist your friends.** Choose peers whose judgment you trust and ask them to read your drafts. Tell them you'd appreciate it if they could let you know both what is working well and why, and what needs to be improved and why. (You can use your peers to help you develop your skills at giving feedback, too, by offering to comment on their work in return.) **In political science, it's important**

to remember that the peer feedback should not be about politics. If your peers are not in your course or are not political science students, be sure to instruct them that their comments shouldn't be about whether they agree with your positions or not. If you have enough time and think your peers would be willing to look at your paper more than once, getting two rounds of feedback is often helpful. The first time, you might ask for help with "big picture" questions about organization and use of evidence; the second time, you might ask for feedback on grammar and style. If English isn't your first language—even if it is your first language—it's often wise to ask a friend to look over your final draft and point out any errors that interfere with the clarity of your prose once you're sure the overall argument is in good shape.

When visiting the writing center or asking friends to review your work, it's often helpful to **bring the assignment prompt and any grading rubrics your instructor may have given you to help your reader get a better sense of what your paper is supposed to accomplish**. Otherwise, if the paper sounds great but isn't actually doing all the right things, your reader will have no way to know that.

Obviously, it's important to make sure that your work is your own; asking someone to tell you where your writing is unclear is different from asking that person to do your writing for you. You may encounter some instructors who tell you not to ask for help from others at any point during the course, and even instructors who generally applaud getting feedback may draw a distinction between papers (where feedback is recommended) and take-home exams (where it may be forbidden). In cases where you've been told not to seek outside help, doing so is a violation of academic integrity and could lead to a variety of penalties. (See the section on plagiarism in Chapter 7 for the kinds of penalties we have in mind.)

It's also important to realize that not all feedback is equally good. If you get suggestions from an instructor, odds are you should take them. But **what if you have two friends read your paper and they give you conflicting advice?** In that case:

1. Trust your own judgment. If one person's advice makes more sense to you than the other's, that's the advice you should take. **But remember, don't equate "making sense" with "agrees with my politics"!**
2. Make sure you don't choose one set of advice over another only because implementing that advice is easier.
3. If you're really not sure, put the paper away for a couple of days and let your subconscious work on it. (We mentioned that it's a good idea to start writing and getting feedback early, right?) Advice that is hard to hear at first often doesn't sound as bad after you've let it sit for a while.

It takes practice to learn what advice you should take and what you should ignore, and it's a good skill to develop because that question never goes away. Academic journal editors sometimes do writers the favor of telling them which reviewer's comments need to be addressed and which don't, but they often leave it up to the writer.

One last thought about feedback: you've probably noticed that you often don't get any from an instructor until it's too late, at least in terms of your current semester's grade. The comments you receive arrive on your final draft, leaving you no opportunity to apply them. We sympathize—you can tell how valuable we think it is to get feedback on drafts. But don't throw those comments out. Just as you can sometimes apply the feedback you receive on the first paper you submit in a course to a later paper, you can also sometimes apply feedback from one course to later courses.

In other words, we recommend that you take the long view and think about feedback through two different lenses:

1. How you can apply it to your current paper or course.
2. How you can apply it to your long-term development as a writer.

Grading Rubrics: A Form of Feedback

We understand that you probably care about grades. You may, in fact, have bought this book to ensure that you'll get good grades in your political science courses. We don't say very much about grades in this book because we think focusing on them is not helpful—you'll do better overall if you think about doing your work well, instead of stressing about what your grade is going to be.

But, of course, most writing assignments are graded. Instructors debate how this grading should be done. Some think that writing, unlike, say, math exams, is so subjective that one shouldn't even try to formulate criteria for its evaluation. Others spell out criteria for what they are looking for but don't say how exactly they translate those criteria into grades. Or they might have exact translations but not share them with you, the students. Or they might offer a highly detailed rubric (Table Appendix A.1).

What to do when your instructor does make his or her criteria or even specific grading rubric available to you? The reason some hesitate is that they worry you will "write to the test"—that is, focus on the criteria so narrowly that you stifle your creativity. Perhaps. But if the criteria or the rubric is really well defined, you *should* write exactly to the specifications. They will teach what your instructor expects from this particular kind of assignment.

Unfortunately, it's hard to come up with good criteria, and even for thoughtful instructors, there might be some distance between what they say they expect and what they think deserves the highest grade when they're reading the papers their students actually turn in.

TABLE APPENDIX A.1 Sample Rubric

	Exceeds Expectations	Meets Expectations	Partly Meets Expectations	Barely Meets Expectations
Interpretation, conceptual analysis, reasoning and logic	All arguments are supported with appropriate evidence. Analyses are interesting and plausible. The analysis and use of concepts are appropriate and convincing. Arguments are sound and valid.	Arguments are supported with evidence. Analyses are plausible. Command of concepts discussed and used reasonable. Arguments are not fallacious.	Confusion between arguments and opinion; inadequate use of reasons and evidence. Command of concepts imprecise but intelligible. Arguments, where offered, may be implausible or fallacious.	Claims supported with minimal or no evidence. Command of concepts confusing or inappropriate. Logic, if used, fallacious.
Structure	The paper states a clear thesis in the introduction. The body of the paper is structured in a way that both the argument and the prose flow. There is adequate signposting for the reader to know what each part of the paper is doing. Each paragraph is structured around a topic sentence. The conclusion offers more than a summary of what has been said.	The paper states a thesis in the introduction. The body of the paper makes sense, although there may be a lack of flow. The conclusion doesn't offer more than a recap of the argument.	The paper has a recognizable structure, but it may be difficult to find the thesis or to see how different parts connect to the overall plan. Some key parts—thesis, introduction, conclusion—may be minimal, confusing, or missing. Paragraphing may be problematic.	The paper offers no thesis. Its different parts do not connect with other parts or make sense on their own. Use of paragraphs, if any, is confusing.

continued

	Exceeds Expectations	**Meets Expectations**	**Partly Meets Expectations**	**Barely Meets Expectations**
Prose style	The prose used is clear and precise. Word choices are appropriate. All sentences communicate clearly to the reader.	The prose used is intelligible. Grammar, word choices, spelling, and punctuation are mostly correct.	The prose used takes some effort to understand. It may be weakened by word choices, over-writing, grammatical errors, or lack of proofreading.	The prose is hard to follow. Grammatical, typographical, and proofreading errors distract the reader.
Formatting	The paper has a title. Its use of citations is as specified in the prompt. It includes page numbers. It uses a reader-friendly typeface.	The paper has a title. It cites the sources used. It includes page numbers.	The paper demonstrates that its author has made some effort toward formatting it.	The paper shows that the author has not thought about formatting.

So if their expectations are made explicit, whether with grade translations or not, you should study them carefully. But don't think of them as a checklist of things to do. Instead, try to understand what kind of paper the instructor wants, and how to write that paper, given those criteria. Then do your best! If your grade isn't what you expected, it's probably better not to tell the instructor, "Hey, I did all of these things! Why didn't I get an A?" It's always good to go over your work with your instructor, but instead of complaining about why you didn't get what you were promised, you really want to think about it in terms of understanding the expectations better for the future.

APPENDIX B

FURTHER INFORMATION ON COLLECTING AND REPRESENTING DATA

Research Guides

Numerous research guides for the social sciences exist. Here are a few different kinds.

- Johnson, Janet Buttolph and H. T. Reynolds. *Political Science Research Methods.* Thousand Oaks, CA: CQ Press, 2012.
- King, Gary, Robert O. Keohane, and Sidney Verba. *Designing Social Inquiry: Scientific Inference in Qualitative Research.* Princeton, NJ: Princeton University Press, 1994.
- Marsh, David and Gerry Stoker, eds. *Theory and Methods in Political Science.* New York: Palgrave Macmillan, 2010.
- Powner, Leanne C. *Empirical Research and Writing: A Political Science Student's Practical Guide.* Thousand Oaks, CA: CQ Press, 2014.
- Van Evera, Stephen. *Guide to Methods for Students of Political Science.* Ithaca, NY: Cornell University Press, 1997.

Visualizing Data

- Few, Stephen. *Show Me the Numbers: Designing Tables and Graphs to Enlighten.* Oakland, CA: Analytics Press, 2004.

- Tufte, Edward R. *Beautiful Evidence*. Cheshire, CT: Graphics Press, 2006.
- Yau, Nathan. *Data Points: Visualization That Means Something*. Indianapolis: Wiley Publishing, 2013.
- Yau, Nathan. *Visualize This: The FlowingData Guide to Design, Visualization, and Statistics*. Indianapolis: Wiley Publishing, 2011.

We also recommend you check out Nathan Yau's website, FlowingData (flowingdata.org).

NOTES

Chapter 1

1. For the first example, see Lee Epstein et al., "The Judicial Common Space," *Journal of Law, Economics, and Organization* 23 (June 1, 2007): 303–325. For the second, see Richard F. Fenno, "US House Members in Their Constituencies: An Exploration," *The American Political Science Review* 71 (1977): 883–917. For the last, see Nicholas A. Valentino, Vincent L. Hutchings, and Ismail K. White, "Cues That Matter: How Political Ads Prime Racial Attitudes During Campaigns," *American Political Science Review* 96 (2002): 75–90.

2. For the first example, see R. Harrison Wagner, "Uncertainty, Rational Learning, and Bargaining in the Cuban Missile Crisis," in *Models of Strategic Choice in Politics*, ed. Peter Ordeshook (Ann Arbor: University of Michigan Press, 1989), 177–205. For the second, see Scott Atran and Robert Axelrod, "Why We Talk To Terrorists," *The New York Times*, June 29, 2010, http://www.nytimes.com/2010/06/30/opinion/30atran.html.

3. John C. Bean, *Engaging Ideas: The Professor's Guide to Integrating Writing, Critical Thinking, and Active Learning in the Classroom*, 2nd Edition (San Francisco: Jossey Bass, 2011), 46–47.

4. The most prominent of such blogs as of this writing is The Monkey Cage (http://www.washingtonpost.com/blogs/monkey-cage/), which the *Washington Post* purchased in 2013. Others include The Duck of Minerva (http://www.whiteoliphaunt.com/duckofminerva/) or the somewhat interdisciplinary Lawyers, Guns, and Money (http://www.lawyersgunsmoneyblog.com/).

Chapter 2

1. Sweetland Center for Writing. "How Do I Make Sure I Understand an Assignment?" Available online: http://www.lsa.umich.edu/sweetland/undergraduate/writingguides/howdoimakesureiunderstandanassignment

Chapter 3

1. This is, in fact, the argument of Robert W. Mickey's *Paths Out of Dixie* (Princeton, NJ: Princeton University Press, 2015).
2. See Nancy Sommers, "Revision Strategies of Student Writers and Experienced Adult Writers," *College Composition and Communication* 31, no. 4 (December 1980): 378.
3. David Stasavage, "Was Weber Right? The Role of Urban Autonomy in Europe's Rise," *American Political Science Review*, 108 (2014): 337–354.
4. Seok-Ju Cho, "Voting Equilibria Under Proportional Representation," *American Political Science Review*, 108.02 (2014): 281–296.
5. Simon Chauchard, "Can Descriptive Representation Change Beliefs about a Stigmatized Group? Evidence from Rural India," *American Political Science Review*, 108.02 (2014): 403–422.
6. Quoted in Barbara Herman, "Editor's Introduction" in John Rawls, *Lectures on the History of Moral Philosophy* (Cambridge, MA: Harvard University Press, 2000), xvi–xvii.

Chapter 4

1. This table is based on the Militarized Interstate Disputes data set, version 4.1, available at http://www.correlatesofwar.org/data-sets/ MIDs. See Faten Ghosn, Glenn Palmer, and Stuart Bremer, "The MID3 Data Set, 1993–2001: Procedures, Coding Rules, and Description." *Conflict Management and Peace Science* 21 (2004): 133–154. The regression table output is created using Marek Hlavac, "stargazer: LaTeX code and ASCII text for well-formatted regression and summary statistics tables" (2014), R package version 5.1. http:// CRAN.R-project.org/package=stargazer.
2. Edward R. Tufte, *Beautiful Evidence* (Cheshire, CT: Graphics Press, 2006).

Chapter 6

1. On disciplinary hedging, see Robert Madigan, Susan Johnson, and Patricia Linton, "The Language of Psychology: APA Style as Epistemology," *American Psychologist*, 50 no. 6 (1995): 428–436.

2. George Gopen and Judith Swan, "The Science of Scientific Writing," *American Scientist*, 78 no. 6 (1990): 550–558. This article is also freely available at http://www.americanscientist.org/issues/pub/the-science-of-scientific-writing/1

Chapter 7

1. You will find it at www.start.umd.edu/gtd.
2. The reference in the block quotation comes from Catharine MacKinnon, "Defining Rape Internationally: A Comment on Akayesu," *Columbia Journal of Transnational Law* 44, no. 3 (2006): 957.
3. Jason Barabas, Jennifer Jerit, William Pollock and Carlisle Rainey, "The Question(s) of Political Knowledge," *American Political Science Review*, 108, no. 4 (2014): 840–855.
4. For a book: Author(s). *Title*. Publisher, year.
5. Jane Smith. *Fake Text Title for Footnote Sample*. (New York: Excellent Publisher, 1998).
6. Jane Smith. *Fake Text Title for Footnote Sample*. (New York: Excellent Publisher, 1998), 14.
7. Ibid.
8. Ibid., 19.
9. Fred Jones. *Another Fake Text Title*. (Boston: So-So Publisher, 1999).
10. Smith, *Fake Text*, 19.
11. For APSA Style: http://www.apsanet.org/portals/54/Files/Publications/APSAStyleManual2006.pdf; for Chicago Style: http://www.chicagomanualofstyle.org/tools_citationguide.html.

INDEX